William Buckler

The Larvæ of the British Butterflies and Moths

Vol. III

William Buckler

The Larvæ of the British Butterflies and Moths
Vol. III

ISBN/EAN: 9783743426146

Manufactured in Europe, USA, Canada, Australia, Japa

Cover: Foto ©berggeist007 / pixelio.de

Manufactured and distributed by brebook publishing software (www.brebook.com)

William Buckler

The Larvæ of the British Butterflies and Moths

.

THE

RAY SOCIETY.

INSTITUTED MDCCCXLIV.

This volume is issued to the Subscribers to the RAY SOCIETY *for the Year 1888.*

LONDON:

MDCCCLXXXIX.

OF THE

BRITISH BUTTERFLIES

AND

MOTHS.

BY

(THE LATE)

WILLIAM BUCKLER,

EDITED BY

H. T. STAINTON, F.R.S.

VOL. III.

(THE CONCLUDING PORTION OF THE BOMBYCES.)

LONDON:

PRINTED FOR THE RAY SOCIETY.

MDCCCLXXXIX.

PRINTED BY ADLARD AND SON, BARTHOLOMEW CLOSE.

PREFACE.

In the second volume of this work the larvæ of the first twenty-seven of our Bombyces appeared; the present volume includes the remainder of our Bombyces.

Since the publication of Vol. II, the Ray Society has sustained a most serious loss in the death of the Rev. John Hellins, who had for so many years worked with Mr. Buckler at the subject of which these volumes treat. Though Mr. Hellins had not been in robust health of late years, it was hoped that he would have been spared to us for some years longer, to assist in the completion of several more of these volumes.

The task which had specially devolved on the Rev. John Hellins was that of writing out the descriptions of those numerous larvæ, of which, although they had been faithfully depicted by Mr. Buckler, no descriptive notes by him were extant amongst his papers, nor had he at any time published descriptions of them. Thus, out of the seventy-nine larvæ figured in the present volume, only thirty-two were described at the time of Mr. Buckler's decease.

We had hoped that descriptions of a large proportion of the forty-seven which were thus deficient would have been furnished by Mr. Hellins, but, unfortunately, his sudden death on the 9th May, 1887, when just in the very midst of his labours (when he was actually receiving larvæ by almost every post from some of his numerous friends) compelled us to seek for help in other quarters.

This, however, we found a most difficult matter; those who had the necessary skill and capacity for describing larvæ lacked the time. After some anxious correspondence on the subject, it was hoped a solution of the difficulty had been found, and it was announced in the pages of the 'Entomologist's Monthly Maga·zine,' vol. xxiv, p. 20, in the obituary notice of the Rev. John Hellins, that "at the earnest solicitation of the Ray Society, Mr. W. H. B. Fletcher, of Fairlawn House, Worthing, Sussex, had kindly undertaken to take up the broken thread of Mr. Hellins' work."

Since this announcement was made in June, 1887, Mr. Fletcher has devoted much time and no small amount of patience to the task, but unfortunately the more he worked at the subject the more it seemed to grow, and his ideas of what a complete description should be developed even in a larger proportion; so that just when his MS. was expected to be ready for press he found that still another year or two of observation was needed to render the matter which he had to furnish worthy of the subject and of the reputation of the Ray Society.

In this dilemma he wrote to me to the following effect :

BERSTED LODGE, BOGNOR, SUSSEX;
December 31st, 1888.

MY DEAR SIR,

I am afraid the object of this letter will not be very agreeable to you, and I am certainly sorry

to have to write it. Briefly, it is to ask you to release
me from the work I have undertaken for the Ray
Society. I find that to do it as it should be done
would require all my time unceasingly, summer and
winter, for many years. This I cannot give to it.
My engagements other than entomological are so
increasingly numerous that were I entirely to give up
collecting on my own account, and to devote myself to
the Ray Society's work alone, I could not carry it out
as it should be done.

When you first asked me to undertake the work, I
thought that Mr. Buckler had written out most of the
life-histories and that merely an odd species here and
there had to be worked out if possible. I little dreamt
that nearly all the common, and many of the less
obtainable larvæ had not been described by him.

Had I a lot of old material by me, the position would
be different from what it is. Coming in while the
volumes are being published, and starting anew with
no chance of checking one year's work by repeating it
in the following season, can only result in the accumu-
lation of a mass of undigested information probably
full of inaccuracy and, in any case, quite unworthy of
the splendid series of volumes published by the
Society.

The following reasons, among others, have made
me realise that it is not possible for me to carry out
the task properly :

(1) During my absence from home, and indeed,
owing to pressure of work when I am there, larvæ
often change their skins and pass a stage unobserved
and undescribed, when, in my opinion, all the labour
spent on them becomes useless, and should be repeated

another year, this being, under the present circumstances, impossible.

(2) I find it is only possible to study a few life-histories in any one year.

(3) It is only in trying to write out some of my notes that I have found out the form in which they can best be taken. Needless to say that mine, copious as they are, are not in that form.

When inviting me to take up the work hitherto done by Mr. Hellins, you introduced the subject by asking the question "Who is to do it?" May I now tardily answer it, Sussex fashion, with another query, "Why do it?" Why not publish the Plates, together with Messrs. Buckler's and Hellins' articles and notes, as their work pure and simple? This course would, in my opinion, result in the production of a grand work of reference, which would be for the United Kingdom what Hübner's great work is for the Continent.

I must again apologise for the inconvenience you have been caused by the course I have taken. You will probably agree with me that to have been inundated with a mass of ill-digested materials would have been a worse evil still, and will allow at least that I am doing at last what I ought to have done earlier in asking you to release me from the duty I had undertaken.

With the best wishes of the season,

I am

Yours very truly

W. H. B. FLETCHER.

H. T. STAINTON, Esq., F.R.S.,
  Mountsfield, Lewisham.

In this position of affairs it has been decided to

restrict the letterpress of this volume to the materials left by Mr. Buckler and Mr. Hellins, without further seeking for any extraneous aid.

As in the previous volumes, the letters W. B. or J. H. after each description indicate whether written by William Buckler or by John Hellins, and the figures immediately following give the date when the description was written for publication, the reference that follows being to the volume and page of the ' Entomologist's Monthly Magazine,' or to Mr. Buckler's Note Books.

We are again indebted to Mr. G. C. Bignell for a list of the parasites bred from the larvæ of which the present volume treats.

H. T. STAINTON.

Mountsfield, Lewisham;
January 25th, 1889.

CONTENTS OF VOL. III.

CLASSIFIED LIST OF THE SPECIES

IN THIS VOLUME.

HETEROCERA.

THE LARVÆ

OF THE

BRITISH MOTHS.

DILOBA CÆRULEOCEPHALA.

Plate XXXVI, fig. 1.

DURING the winter of 1881–82 the Rev. J. Hellins sent me three eggs of *cæruleocephala ;* they were brown and apparently ribbed, but the ribs could not be counted as they were so curiously covered with bristly brown hairs. On the 6th of February Mr. Hellins asked to have them returned, as he had lost those he had kept for himself, so I returned them. On the 11th of March he reported that two of them had hatched and were bristly little fellows.

On the 30th of March I received one of the larvæ, that had passed its first moult, feeding on whitethorn ; it was 4 lines long or 11 mm., of stoutish and uniform proportion, of a lightish grey colour with yellow dorsal stripe and spiracular line, black head, and a streak of yellow across the upper lip, a black oblong squarish spot across the middle of the second segment, a transverse series of round black dots on the third and fourth, on the back of those beyond they occurred in trapezoids, a larger oval black spot was on the front part of the thirteenth segment, and a black plate on the anal flap ; one black dot on the side of

each segment, a smaller one following rather lower
down near the division and touching the spiracular
yellow line; the black spiracle being situated between
the two dots; close beneath the yellow line occurred
a black dot followed by another lower down and more
behind, and another on each proleg; the belly was
light greyish-green; a black hair proceeded from each
black spot.   Skin smooth and rather glistening.

On the 5th of April, this larva having moulted a
second time, came again to me, and on the 6th I
figured it.   It now measured 6 lines, or 13 mm. long,
and of stouter proportion; the ground colour of the
body a deep bluish-grey, the dorsal and spiracular
stripes bright yellow, the dorsal widening at the end
of each segment, except on the third segment, where it
was a transverse bar in the middle; no yellow on the
head nor on the back of the second segment, where
the ground was of a paler grey than the rest of the
body, as it also was on the thirteenth segment; the
yellow dorsal mark on the twelfth segment was very
much broader than on others; the yellow spiracular
stripe was straight at its lower margin and widest at
the end of each segment; the lobes of the head black
and glossy, centre part of face between them light grey,
marked above the lip with black; mouth black.   The
squarish black plate on the second segment was composed of four black shining spots run together, but
dorsally divided by a fine thread of light grey.   The
other round black spots of the body appeared velvety
from their being covered with short bristly pubescence,
and each spot emitted a black hair.   Belly leaden-grey
with black spots, some of them of smaller size.

On the 13th of April, after the third moult (which
is supposed to have occurred on the 10th—11th), the
length was 15 mm., or nearly five-eighths of an inch,
and of stouter proportion, having an elevated transverse ridge on the middle of the back of the third
segment and another on the twelfth; the ground colour
of the head was greyish-white, likewise the dorsal line

of the second segment and the anal flap ; the colour
of the body, *i. e.* the back and belly, was very dark
slaty-grey, the dorsal stripe and the spiracular stripe
of bright pale yellow as before ; the lobes of the head
had each a large black spot on the crown ; the ocelli
in a black spot below, the face and lip greyish-white,
also papillæ tipped with black, edge of lip black ; the
second segment pale yellow across the middle with
two pairs of black spots in front and again at the back ;
the ridge on the third segment was bright yellow,
bearing two black tubercles, that on the twelfth segment also yellow, bearing four black tubercles in pairs ;
all the tubercles were velvety-black and bristly ; the
dorsal ones on the third and fourth segments bore two
long black hairs, but all the others only one black hair ;
a black, horny oval spot was on the outer side of each
ventral proleg ; anterior legs black. By the 16th of
April it had become 9½ lines long or 20 mm., and by
the 18th was 11 lines or 23 mm. long.

By the morning of April 25th it had moulted again
for the fourth time, and on the 29th measured 1¼ inches
or 32 mm., stout and cylindrical ; a transverse elevated
ridge across the middle of the third and twelfth segments of bright pale yellow ; the dorsal yellow stripe
was a little interrupted at the end of each segment ;
the spiracular stripe was broadish and abruptly contrasted with the dingy greenish smoky-grey of the
belly, but above was softened off a little into the
greenish-grey of the side, and this was separated from
the very pale bluish-grey of the back by a subdorsal
stout line of white, very softly edged and faint.

By the 6th of May, when stretched out it was 1½
inches long and stout in proportion, the skin smooth
(apparently) but without gloss, the black spots round,
dull, and rough, each surrounded by a whitish halo.
By the 8th it was full-fed, and in a day or two began
to contract in length and wander about until the 12th,
when in the course of the night it spun itself up within
an oval cocoon of whitish silk overlaid with portions

of hawthorn leaves, so as to cover and almost hide the cocoon. The moth, a female, emerged on the 26th of October.

The oblong cocoon when cut open was found to be very tough and strong, white and smooth within. The pupa of true *Bombyx* shape measured 7 lines in length and of stout proportions, the abdominal tip furnished with two lateral projections suggestive of the previous anal prolegs, and each bearing several bristly hairs, and traces yet remained, though minute, on other parts of the former hairs of the tubercular warts; the entire surface of every part being very dull and rather rough, while a band of stronger roughness, approaching almost to thorny points, was on the middle of the back of each abdominal segment. The colour was of a sooty brownish-black or blackish-brown; spiracles rather prominent. (W. B., Note Book IV, 104.)

PETASIA NUBECULOSA.

Plate XXXVI, fig. 3.

On the 6th of May, 1881, I received from Mr. H. McArthur, while he was collecting at Rannoch, a dozen eggs of this species, laid loose or on small morsels of bark; of these two proved infertile, the first egg hatched on May 16th being one I had previously sent to the Rev. J. Hellins; with me two were seen to be hatched in the early morning of the 17th, two at midnight, three by next morning, one near midnight following, and the last one by the morning of the 19th. All my young larvæ took to birch readily, but the one in Mr. Hellins' care chose oak, and fed on it until its third moult, and from that time, the 4th of June, it would eat birch and not oak. After feeding their growth was very perceptible, and when six days old they each in turn lay up for moulting; this operation occurred five times in all before their full growth was attained.

Generally they agreed very well together, though two individuals during the earlier stages, while helplessly laid up waiting to moult, appeared to have been inconveniently in the way of some of the others, and so got fatally bitten behind; afterwards, with more space, they proved to be very contented and well-behaved.

They became full-fed from June 26th to 29th and retired to earth; over the earth, at the end of June, I placed a thick covering of moss, and found afterwards that only two had elected to remain below in the earth, and the other five were lying in the pupa state on its surface beneath the moss; the larva, with Mr. Hellins, had buried itself four or five inches deep in the loose leaf-mould furnished for its retreat.

I bred three male moths and one female in March, 1882; the single pupa of Mr. Hellins' stood over a second winter and disclosed a fine male imago, February 15th, 1883 ; my remaining pupæ produced five male and female specimens April 1st, 1883.

It has been pointed out before that the egg of *Nubeculosa* (as well as those of *P. Cassinea* and *Diloba cœruleocephala*) does not so much follow the *Notodonta* as the *Noctua** type, being circular and convex above, with a largish central space covered with irregular reticulation, and on the sides from forty to forty-five blunt ribs, with somewhat coarse transverse lines ; in height about one thirty-sixth of an inch, in width about one twenty-fourth ; the shell rather glistening, the colour at first dirty drab-green but soon becoming closely and tortuously streaked and blotched with blackish-green ; a few hours before hatching these marks become indistinct and clouded, and the shell looks somewhat shrivelled.

The newly-hatched larva is about one tenth of an inch in length, with the first and second pairs of ventral prolegs less developed than the third and fourth pairs, so that the walk is semi-looping ; the head of a rather light, shining orange-brown colour ; the back slaty-grey ; the

* Many systematists class these species amongst the *Noctuæ*—H. T. S.

sides pale drab ; the black warts very large and round, each furnished with a small black bristle. In this stage the likeness to *cæruleocephala* is marked, but at each moult the warts become proportionately smaller and less conspicuous, besides assuming another colour, and so this resemblance disappears. From the first the young larva eats small holes quite through the leaves of its food, and I noticed its habit of spinning a few threads for a foot-hold.

After the first moult a slight protuberance appeared on the twelfth segment and front portion of the thirteenth ; the ground colour was pale greenish, bearing dorsal to subdorsal lines of paler dots, and on the middle segments a wide sort of incomplete V in very fine black lines ; the black tubercular dots were much smaller than before, and only to be seen with a lens, but their bristles had become longer ; the anterior legs were black, and on the outside of each ventral proleg was a black spot.

After the second moult the head was pale shining green, the body light dull green, having a purplish tinge in it, the tubercular dots pale yellowish, the dorsal markings composed of elongate whitish-yellow dots, two on a segment, and along the subdorsal region were four yellowish dots on each segment, a slanting streak of the same colour appeared on the side of the fourth, and a transverse streak on the ridge of the twelfth, and a black spot on each ventral proleg as before.

Having moulted the third time, June 3rd—5th, the larvæ began to assume their well-known star-gazing posture, with all the front part of the body extended upward in a curve, bringing the head so far back as to be elevated just over the eleventh segment, while the anterior legs were freely outspread, the third pair wider apart than the others ; all the details of colour being similar to those of the previous stage.

The fourth moult happened on the 9th—10th of June, and they soon resumed feeding, eating large pieces out of the leaves at intervals, and at other times were to be seen for long periods hanging to the birch sprays motionless in their singular attitude of repose, but yet so suggestive of great muscular exertion and watchfulness. Their growth now seemed rapid, as in the course of three days they were observed, when in motion, to be an inch and three lines long, stout, and thickest behind, their colouring of the same light green as before, the upper surface bearing rather warty spots of bright yellow and slash-like streaks of the same yellow on the thoracic and posterior segments ; the anterior legs black, ringed with ochreous at the joints. Some individuals still bore the large roundish black spot above the foot of each ventral proleg, while others had only a black outline of it, or part of it.

The fifth moult occurred between the 15th and 19th of June, and for a time after this operation the head was of rougher texture than heretofore, but gradually in three or four days it regained its glossiness. The larva did not now so often assume its posture of contemplative repose, but seemed more intent on its consumption of food, and in the shorter intervals of rest was to be seen lying quite at full length, or in a gentle curve, along the birch twigs, quite flat and lethargic, until almost full-fed ; but when this stage was reached, it was again frequently to be seen in its more characteristic position. When quite full-grown the larva was 2 inches in length and of thickness in proportion, with a very soft skin ; the head full and rounded, with lobes slightly defined ; the body cylindrical, with plump segments deeply divided as far as the twelfth, and there tumid and humped with a slight dorsal ridge, then sloping, and tapering a little on the very long front part of the thirteenth and still more on the short anal flap, deep wrinkles subdividing only the thoracic third and fourth segments. The anterior legs rather small, but set on large pectoral muscular foundations ; the ventral and anal prolegs stout, with well-developed feet, and hooks to secure prehension and progression.

The colour of the head was now pale bluish-green, the
upper lip whitish or else pale yellow, the mouth black,
the back of a delicate pale yellowish-green, becoming
paler and opaque from the thoracic segments to the
twelfth, and blending gradually into a deeper brilliant
yellowish transparent green on the sides and belly.
The slightly raised spots were all of pale primrose-
yellow, the dorsal series elongate-oval in shape, two
on each segment, one beyond the other, in a broken
line on the fifth to the eleventh inclusive; the other
series of spots were of round shape, such as the trape-
zoidally-arranged fours of the back, the subdorsal
broken line of threes, the lateral single spot, and the
single spot below each spiracle, which was itself white,
tenderly outlined with black; a transverse series of
four spots showed faintly on the fourth segment, a
small tumid side streak of the same yellow was on the
third, and another conspicuously larger and longer
was on the fourth, slanting down obliquely forwards;
two spots were on the back of the twelfth segment,
and behind them on the summit two much larger spots
united to a tumid curved streak of yellow; a conspicuous tumid side streak of similar yellow began
behind the spiracle, tapering off on the margin of the
anal flap. The anterior legs were bright red, and outside each ventral proleg was a roundish ring of black,
the feet being furnished with brown hooks.

The pupa is a full inch in length, and 4½ lines
in width at the thickest part across the ends of
the short wing-covers, the antenna-cases well developed; the head and thorax smooth, the wing-covers
most minutely roughened, also the upper portions of
the abdominal rings; the free segments of the abdomen
are very deeply cut, and taper gradually towards the
end, but with dissimilar outline on the ventral and
dorsal surfaces; the ventral becoming bluntly rounded,
and the dorsal rising somewhat in a hump, from which
springs the base of a prolonged stout spike, whose
blunt extremity is furnished with two fine tapering

points bent downwards and curved like claws; the colour is a deep and dingy red during the first year, and in the second becomes a blackish-brown, bearing a slight purplish gloss. (W. B., 9, 4, 83; E.M.M. XIX, 271.)

PERIDEA TREPIDA.

Plate XXXVI, fig. 4.

On the 26th of April, 1870, Mrs. Hutchinson kindly sent me twelve eggs of *trepida*, which began to hatch May 11th, and were all out on the 12th.

The egg is circular, convex above, flattened beneath, smooth; of a delicate bluish opaque white, which it retains to the last; the only slight change that occurs just a few days before hatching is that a small grey speck becomes indistinctly visible in the centre.

The newly-hatched larvæ were robust-looking little fellows, of a greenish-yellow colour, with black dots and hairs, the head and its lobes being outlined with black. At this early age, as soon as extruded from the egg, it assumes the posture in repose, which is so characteristic of this species through all its larval existence, its back forming a hollow curve with the head and tail erected free from the surface of the leaf to which it is attached.

At its third moult its length was about five-eighths of an inch, it had then its characteristic stripes, viz. a double pale yellow dorsal and oblique side stripes on a green ground colour, from which the black dots had disappeared. * * * * (W. B., Note Book II, 136.)

On the 23rd of May, 1882, I received two eggs laid on the *underside* of oak bark, set up in a wood to dry about a week before; these eggs were part of a batch so found by a son of Mr. W. R. Jeffrey, who sent them to me.

The shape of the egg is hemispherical, that is,

rounded above and with a flattened base attached to the bark, apparently smooth and of pure white surface, though when they came I could just see a faint light brownish spot showing through the top of the egg, the shell being otherwise quite opaque. They both hatched in the early morning of the 27th, and the shell could then be seen to be quite thick, of a bluish-green substance within, and externally with a layer of opaque white.

The head of the young larva was remarkably large, the body tapering thence behind; in colour it was wholly of a light, rather olive- or ochreous-greenish; anterior legs black and dots blackish, each dot having a fine black hair. On the night of June 2nd they moulted the first time, and by the morning of the 3rd they were feeding quite through the leaf from the edge (previously they had eaten between the veins, skeletonising the margin of the leaf). They were green in colour, with a streak of blackish behind, down each cheek to the mouth, the back rather deeper green with a darker dorsal line, and a faint yellowish subdorsal line; dots and hairs black.

On the 10th and 12th of June they moulted the second time, and in four days the slanting yellow streaks appeared on the sides as puffed slashes; the double dorsal, pale yellow lines, having between them a dark green central line, were suggestive of the future design, the subdorsal line yellow and very thin, the slanting side stripes faintly edged with dark red; on the head was a fine black streak down the middle of each lobe, and another down the back of each cheek.

Both larvæ moulted the third time on the 21st, and both for the fourth time on the 30th of June, and fed well the next day, but on the 3rd of July I found one was lying dead. The other became a fine thick fellow, brilliantly coloured, but by the 14th July it was becoming of a more dingy green, and the next day had spun itself up in a brownish cocoon between leaves of oak. (W. B., Note Book IV, 112.)

ORGYIA ANTIQUA.

Plate XXXIX, fig. 1.

Eggs laid on a cocoon-like web upon a spray of *Acacia dealbata*, in the gardens of the Crystal Palace at Sydenham, sent me by Mr. George Thomson, March 18th, 1879.

The eggs laid close and evenly together side by side. The egg is rounded, having near the top a slight rim, which swells out a little, the flat top having a central depression. Colour of the egg shining reddish-brown. These eggs never changed colour. They began to hatch June 21st, 1879, a few at a time.

The newly-hatched larva was dark brown, with segments 5 to 8 darker brown, a paler triangular spot on the back of the ninth, a darker brown spot on the twelfth; the tubercles dark brown, one on either side of the front of the second segment longer than the others, with longish brown hairs, some longer than others.

After its first moult it grew to a length of three-sixteenths of an inch, it had then a pale patch on the back of the fourth segment, which rather divided in front, encroaching a little on the third segment; another distinct pale patch was on the back of the ninth, and a palish dorsal spot on the back of the tenth and eleventh, and a transverse pale mark on the front of the thirteenth segment; a pale but rather interrupted subdorsal line was visible from the fifth, faintly to the ninth, and thence to the twelfth segment distinctly.

By the 19th of July a great many had completed their third moult, and had now assumed the character-istic tufts of hairs like shaving brushes on the back, on the fifth and sixth segments blackish, on the seventh and eighth whitish, and on the tenth and eleventh a pinkish tubercle, a longish tuft of black clubbed hairs on the twelfth, slanting backwards; on each side of the front of the second segment, at an angle laterally pointing forwards, a tuft or fascicle of longer hairs,

but of varying lengths, blackish-plumed and clubbed at
their tips.

By the 27th three or four had accomplished their
fourth moult, with tufts of quite white hairs down their
back; they were then three-quarters of an inch long.

By the 19th of August they were full-grown, and
1¼ inches in length; the white brush-like hairs on
the back tipped with brownish, the black hairs plumed
and clubbed as before on each side of the second
segment, and on the back of the twelfth; cream-
coloured hairs radiated from the other tubercles, longer
and more numerous on the lower rows along the sides;
ground colour of the body cool velvety-grey, a black
velvet stripe down the back, more or less interrupted
by the brushes and by brilliant red tubercles, a black
or blackish line followed on each side more or less
distinct or broken; head shining black, upper lip pale
creamy-white, mouth black; on segments 9, 10, 11 and
12 the black dorsal stripe was bordered with cream
colour as a narrow stripe.

In some individuals the ground colour was greyish-
olive, and the red tubercles were outlined with cream
colour.

All the above died off nearly mature.

I received some more eggs on the 9th April, 1880;
they began to hatch May 14th.* I fed the young
larvæ with sallow. The first moult took place the 18th
and 19th of May. These larvæ throve well on sallow,
and the earliest began to spin their cocoons on the
17th and 18th of June. The moths began to appear
on July 10th, when I bred two ♂ and two ♀; on the
13th appeared three ♂ and on the 17th one ♂. (W.
B., Note Book IV, 1.)

* The larvæ continued to hatch about two a day, sometimes three,
but generally two until the 29th May, when there were twenty-one
hatched, and I cast adrift the remaining eggs.

MILTOCHRISTA MINIATA.

Plate XL, fig. 3.

Eggs were obtained from a female which had been captured July 18th, 1867. The larvæ were hatched before the end of the month, they fed slowly but almost continuously till the end of the following May, by which time six out of nineteen survived to spin up. The moths emerged between the 19th and 30th June, 1868.

The food at first chosen was a sallow leaf, which had become damp and rotten by being kept in a glass-stoppered bottle; afterwards when placed out of doors in a flower-pot they ate withered oak and sallow leaves and various lichens; in the spring they nibbled the slices of turnips put in with them as traps for slugs, and at last settled down steadily to eat the red waxy tips of *Lichen caninus*, and fed up to quite full size on this food. In a state of nature I understand they are found feeding upon the lichens that grow on the boles of oak trees.

The eggs of *miniata* are very different from the usual round pearly beads of the *Lithosiæ*, being more fusiform in shape, rich yellow in colour, and placed on end with great regularity at a little distance from each other in rank and file. My batch of eggs was deposited in four rows, viz. three of five eggs each, and one of four.

The larvæ from the first were little dingy, foggy-looking fellows, with a quantity of fine hair on their backs, and although after the last moult their plumes became denser and darker than before, yet a description of the last stage is applicable throughout.

When full-grown, the length is a trifle over half an inch, the hairs that project before and behind making it look a little longer; the figure stout, uniform in bulk; the skin very shining, but densely covered with plumes. Segments 2 and 13 are furnished only with

very short simple hairs, but the other segments have
each six whorls of wonderful plumose verticillate
hairs, those on segments 3 to 7 being fully one-eighth
of an inch high, and those on segments 8 to 12 a little
shorter, while along the sides and just above the feet
are tufts of plain hairs. When looking at one of them
in motion, I could not help mentally comparing it to
an animated hearse with palish plumes.

The colour of the skin, when it can be seen, is a
waxy dark drab; the plumes from the head to segment
7 are blackish mouse colour, and the rest a paler tint
of the same. When disturbed the larva bends into a
circle, placing the two extremities together, with the
tufts standing out apart.

The cocoon is a long-oval in shape, very slight but
close in texture, the silk wonderfully interwoven with
the cast-off plumes stuck upright, so that whilst fresh
and uninjured by rain it might at first sight be mis-
taken for the larva; one which I watched in progress
was completely finished, so far as outward appearance
went, in twenty-four hours. The pupa is short,
reddish-brown in colour, the cast larva skin adhering
to the anal segments. (J. H., 5, 9, 68; E.M.M. V,
111.)

LITHOSIA CANIOLA.

Plate XL, fig. 4.

A larva, feeding on olive-green house-top lichens,
with a taste for clover, was secured for figuring by the
kindness of Dr. Knaggs, on May 30th, 1862.

Its head was dark brown, the body tapered a little
at either extremity, the ground colour brown, a thin
blackish dorsal line slightly widening in the middle of
each segment, the subdorsal lines composed of cunei-
form orange-red marks pointing backwards, and bordered laterally with similar marks of black, a whitish
spot almost touching the point of each wedge; the sides

rather paler than the back, with a dusky lateral line;
the tubercles studded with brown hairs. (W. B.;
E.M.M. I, 49.)

LITHOSIA AUREOLA.

Plate XL, fig. 5.

The larva was received on August 19th, feeding
upon lichens attached to oak.

This larva was very active in its habits; it was not
yet mature, being but little more than five-eighths of an
inch in length, rather slender, and of nearly uniform
thickness, but tapering very little posteriorly. The
tubercles all tufted.

The ground colour of the back was white, but this
appeared only as four white lines separating the black
dorsal, intermediate and broader subdorsal stripes;
and this pattern was interrupted at the fourth, eighth,
and twelfth segments by dark brownish-black patches
covering the back, and on the fourth and twelfth
looking almost like humps from the greater denseness
of the tufts of hair; and on the ninth segment the
dorsal stripes were absent, leaving the whole area as a
conspicuous whitish spot; the sides, belly, and legs
were brownish-grey; the folds between segments 3
and 4 white; there was a white spot just above the
legs on the third, and a white blotchy line similarly
placed on the fourth. The second segment was dark
brown, with a reddish margin in front, and a short
longitudinal streak from it of the same tint on the
subdorsal region. The dorsal tubercles of all but
the three dark segments were orange-red, bearing
brownish-grey hairs, the first of each dorsal pair
being small in size and the second behind very large,
so as to project beyond the subdorsal stripe, on which
they were placed into the side, and behind each tubercle
of this pair was a white dot; along the sides were
two rows of similar tubercles, the lowest being just

above the legs, thickly furnished with brownish-grey
hairs; a few hairs longer than the rest proceeded
from the thoracic and anal segments; the head itself
blackish-brown.

This species spins up in the autumn and passes the
winter in the pupa state. (J. H., 5, 9, 68; E.M.M.
V, 113.)

LITHOSIA HELVOLA.

Plate XL, fig. 6.

On the 13th June, 1868, I received from Mr. Machin
four larvæ of this species, then not far from full-grown;
their food was a large coarse lichen growing on the
bark of yew trees. In a few days they had spun rather
loose cocoons, with a few grains of earth attached to
the silk, on the underside of the pieces of bark. The
moths appeared July 2nd to 6th.

When full-grown, the larva is nearly three-quarters
of an inch in length, moderately stout, with the
posterior segments tapering slightly towards the tail;
all the tubercles furnished with tufts of hair.

The ground colour of the back varies—being pale
grey, whitish-grey, or white, and the colour of the
sides and belly is grey, brownish-grey, or greenish-
grey; there is a subdorsal stripe of black, separating
the white back from the grey sides, and itself interrupted by one of the hinder pair of tubercles on the
back of each segment; down the centre of the back run
two black lines, which represent the dorsal stripe,
appearing united at the hinder end of all the segments,
as well as on the front of all, except the last four, and
interrupted through the middle of the others; and
and between these lines and the subdorsal stripe comes
another fine black line on the hinder half of each segment. On the fourth segment the space between the
dorsal lines is filled up with black, forming a conspicuous lozenge-shaped mark; on the eighth segment

is another black mark, but triangular in outline; and on the ninth segment the subdorsal black stripe is interrupted by a white spot, which extends somewhat into the grey colour of the side, and along the side run two dark brownish interrupted lines; the head is dark brownish-grey, lobed and freckled with black; the tubercles are grey or brownish-grey, and the tufts of hair growing from them are of the same tint. (J. H., 5, 9, 68; E.M.M. V, 112.)

LITHOSIA STRAMINEOLA.

Plate XLI, fig. 1.

This insect, as previously recorded in the 'Zoologist,' M. Guenée has pronounced to be a variety of *L. griseola* after comparing the figure of the larva with preserved skins of *griseola* in his possession. The larva was depicted June 24th, and the imago appeared July 30th.

The larva was brown, the head of a darker brown, the back of the second, third, and anal segments orange-red, as though the subdorsal marks had become confluent; a similar red mark, of an irregular trapezoidal figure, formed the subdorsal line on the anterior two-thirds of each segment, a thin blackish line bordering them externally, a thin dark brown dorsal line, interrupted on the second and third, and terminating on the twelfth segment, tubercles and hairs brown. (W. B.; E.M.M. I, 49.)

From eggs of *L. stramineola* kindly sent to me in August last by Mr. C. G. Barrett, I have lately succeeded in rearing four perfect insects, one male and three females. Two of the females were yellow all over, one of them having its wing *somewhat clouded with grey;* and the male was *grey all over,* in fact, a true *griseola.*

The correctness, therefore, of M. Guenée's opinion as to the identity of these two forms is completely established, and *stramineola* must take the position

18 LITHOSIA STRAMINEOLA.

which he assigns to it, of being a variety of *griseola*.
(J. H., 14, 7, 73 ; E.M.M. X, 69.)

Two larvæ which had been reared from eggs, were
received from the Rev. J. Hellins, April 29th, 1873,
feeding on lichens ; the smaller of the two sickening
for its final moult. This larva was three-quarters of
an inch long, rather stout in proportion ; the ground
colour *dark* slaty pinkish-grey, very faintly marked
along the sides with paler ; the subdorsal marks of
orange-ochreous were in front of each segment somewhat of a triangular form, with the angles rounded off,
pointing forwards ; from these a thin faint streak of
greyish-ochreous ran backwards to the end of the
segment, but scarcely to be noticed till beyond the
second tubercle, both tubercles interrupting it, the
second much the largest ; the dorsal stripe broad,
faintly darker than the ground, which was itself blackish,
but it was dull black at the beginning, and there was a
black blotch on the side immediately in contact with
the orange-ochreous mark, so that these marks looked
as if on a short transverse black velvety band ; on the
back of the third segment was a conspicuous patch of
orange-ochreous, another on the front division of the
thirteenth segment ; the former partly divided by a
central black line, a little in front and more behind,
where it bore a small blackish-brown hairy tubercle,
on each side and more in front it was bounded by a
large dark tubercle, which gave a lichen-like character
to the form and colour of this patch. The same idea
was suggested by the hind patch of this ochreous colour,
which was only partly divided by a black central line ;
along the side there was an irregular stripe of a greyish
flesh-colour, interrupted by the row of sub-spiracular
tubercles, which was the middle of three rows along the
side, the lowest row being almost on the upper part of
the prolegs, and almost on the belly on the other
segments. The head black and brilliantly polished,
all the rest dull, either of a waxen or velvety appearance ; all the ventral prolegs well developed, paler than

the rest of the skin, and shining semi-transparent; the anterior legs similar in colour and texture. The colour of the hairs of a whity-brown, mixed with black.

On the 4th of May the moult of this larva was complete, and on the 5th it had eaten up the cast skin entirely, hairs and all; the following day it was feeding freely on *Lichen caninus*. The imago appeared June 21st, 1873. (W. B., Note Book II, 7.)

Lithosia complana.

Plate XLI, fig. 2.

I have also been indebted to Mr. Doubleday for a specimen of this larva, which throve well on lichens off fir-trees, and was nearly full-fed June 9th, 1862; the perfect insect appeared at the end of July. The colour of this larva was brown, with a very dark brown head and dorsal line. The subdorsal markings consisted of oblong, somewhat reniform, dull orange-red marks, one on the anterior half of each segment, followed by an interval of the ground colour, and succeeded by a whitish spot; the usual tubercles and hairs dark brown. (W. B.; E.M.M. I, 49.)

The larva of this species has long been known, and descriptions of it have been published by many entomologists; our object, therefore, in introducing any remarks upon it in this paper, is not so much to describe it over again, as to say something about it with reference to the larva of *L. molybdeola*.

Under the latter species will be found an account of two larvæ reared from the egg in 1867–68, of which very careful figures were also taken, with the view of using them for comparison when the larva of *complana* could be procured. And in this way they have been used both this last summer and the summer before, and the following particulars have been noted.

In several points there exists between the larvæ of

complana and *molybdeola* the similarity which is also
shown by their imagos; *complana* is rather the larger
of the two, but there is in both the same figure, the
same arrangement of tubercles, the same sort of hairs
in the tufts ; in their colouring there is the same ground
of dead blackish-grey, the brown tubercles and hairs,
the velvety-black dorsal and lateral stripes, and the
subdorsal row of parti-coloured orange-red and white
spots.

Now, in the descriptions of *complana* we find these
spots called *oval;* " taches ovales " Guenée calls them ;*
"taches arrondies ou un peu ovalaires,"† says
Boisduval; and, as far as we can gather from our
friends who are accustomed to take the larva of *complana* in this country, they do not know of any other
shape for these spots but *oval* or *roundish;* in the two
larvæ of *molybdeola* mentioned above, these subdorsal
spots had no roundness whatever in their shape, but
were *narrowish, oblong, somewhat wedge-shaped* marks.

Boisduval, in his account of *complana*, goes on to
say, " Elle varie un peu pour la couleur et pour la
forme des taches orangées; quelquefois celles-ci sont
blanches sur tous leurs bords avec le centre orangé ;
d'autresfois il n'y a que la partie postérieure de chaque
qui soit orangée. Souvent elles sont alongées ou un
peu triangulaires, et semblent presque former, lorsque
la chenille est en repos, deux raies non interrompues;"
so that we must either give up the shape of these subdorsal spots as a point of difference, or else suppose
that Boisduval had seen larvæ of *molybdeola* as well
as of *complana*. In coming lower down the side, below
the black lateral stripe, which comes next to the subdorsal spots, we reach another point ; and here Boisduval fails us, for he says nothing of the side of *complana*, only that " les stigmates sont peu apparents,"
and " le dessous du corps est grisâtre," and then he
gives the colour of the legs. Guenée is much more

* 'Annales de la Société Entomologique de France,' 1861, p. 50.
† 'Collection Iconographie et Historique des Chenilles.'

precise, "La région latérale est plus pâle" (than the ground colour), "avec des linéaments noirs, marqués, à la place de la stigmatale, de traits fauves, isolés, très fins ;" and other descriptions also speak of a reddish-yellow line running just above the feet. Now, the *description* of *molybdeola* (before referred to) does not help us much here, for it omits some particulars, the importance of which was not then seen ; but the *figures* show most distinctly that, while in *complana* the spiracular region is occupied by *one broader* rust-coloured line, in *molybdeola* there is first a *fine line of pale grey*, then a line of the ground colour, and then a *narrower* line of the rust colour ; and unless the inspection of a larger number of larvæ of *molybdeola* can prove that this arrangement of lines is not permanent, we have in it a good distinctive character ; and perhaps anyone who could place the living larvæ side by side for comparison, would on a careful examination, find others equally good. (W. B. and J. H., 9, 12, 71 ; E.M.M. VIII, 174.)

LITHOSIA COMPLANULA.

Plate XLI, fig. 3.

Said to feed on lichens, though I have not found this the case with the few I have reared ; the first I had fed on oak ; others were taken on buckthorn and dogwood, and this season one on Clematis.

The larva is of nearly uniform thickness ; its colour above is a very dark bluish-grey ; the head, plate on the second segment, broad dorsal line and subdorsal lines black ; the body furnished with black tubercles and hairs, excepting an orange, lateral stripe beginning at the fifth and ending on the twelfth segments, which encloses the spiracles and extends to the prolegs ; the tubercles and hairs on the latter being also orange colour. (W. B. ; E.M.M. I, 49.)

I will only remark that the larva of this species
assumes its lateral reddish-orange stripe at its first
or second moult, when but little over a line in length;
also that it seems to feed and grow more slowly than
the other species.   (J. H., 5, 9, 68; E.M.M. V, 111.)

LITHOSIA MOLYBDEOLA.

Plate XLI, fig. 4.

Mr. Doubleday most kindly transmitted to me some
eggs he had received of this species, and by the time
the parcel reached me (July 26th, 1867) the young
larvæ had appeared. Most of the brood must have
soon perished, but the three which lived till September
were then about half an inch long, and the two final
survivors spun up before the end of May, and appeared
as moths on July 3rd and 4th, 1868.

I could never see that they ate any food I gave them
freely, but at different times I saw that they had eaten
a little of various lichens from trees or banks, wall
moss, withered sallow and oak leaves, slices of turnip
and carrot, knot-grass, and they must have thriven as
well as they would have if they had been at large, for
the two bred moths were not at all smaller than
captured specimens.

I noticed, not in this species only, but in all the
Lithosidæ larvæ I had, that the characteristic markings
and tints were assumed very early—long before they
had attained a quarter of their growth. When full-
grown this larva is rather more than three-quarters of
an inch in length, moderately stout, uniform in bulk;
head very hard and shining; all the tubercles crowned
with tufts of short hairs, mixed with a few longer
ones; of the dorsal tubercles the front pair are small,
and the hinder pair very large.

The ground colour, when seen between the tufts of
hair, is a dead blackish-grey; but the segmental folds

are black; there is a rich velvety, very black, dorsal
stripe; the subdorsal line, being broken on each
segment by the hinder tubercle with its tuft of hair,
must be rather called a *row* of elongated, parti-
coloured spots, each beginning on the hinder part of
a segment, and continued across the fold into the
next segment, until stopped by the tubercle; the
colours being white for about half the spot, and the
tint of a robin's red breast for the remainder, but
owing to the position of the white portion so near the
segmental fold, only the red hinder part of the spot is
to be seen except when the larva is stretched out in
walking. On segments 2 to 4 these spots are altogether whitish. Immediately below comes another
velvety black stripe, broadest at the centre of the
body, and tapering considerably towards the head, but
less so towards the tail; just above the feet comes a
greyish-ochreous interrupted stripe, edged on both
sides with a dark brown line; the tubercles and short
hairs are brown, the longer ones black.

The pupa stout, reddish-brown in colour; enclosed
in a very slight web of silk, under cover of a stone or
piece of moss. (J. H., 5, 9, 68; E.M.M. V, 109.)

LITHOSIA GRISEOLA.

Plate XLI, fig. 5.

Mr. Doubleday kindly sent me eggs on the 11th
August, 1867, from which the larvæ hatched on the
15th of August. By the end of November the larvæ
were nearly half an inch in length and were full-grown
during May. The moths appeared from June 14th to
27th, 1868.

The larvæ fed at first on withered leaves, especially
delighting to riddle decaying sallow leaves full of holes;
but I saw them also eat a little clover, knot-grass, and
various lichens and mosses. Early in the spring they

attacked vigorously some slices of turnip, but afterwards, on attaining some size, they fed away steadily
on *Lichen caninus*, which I have since learnt had been
noticed to occur where the moth is most abundant,
and no doubt forms part of the natural food of the
larva. When full grown the length is quite an inch,
the figure stout and uniform; the head small; all the
tubercles tufted with stiff hairs, which are short on
the back and longer on the sides, with a few of extra
length on the second and thirteenth segments.

The colour is a rich velvety blackish tint above,
dingy blackish-brown below; the central portion of
the back can, however, be distinguished as a stripe of
more intense black than the rest; there is a subdorsal
orange-ochreous stripe, which being interrupted by
the tubercles appears on segments 4 to 12 as a row of
wedge-shaped marks; but on the second segment
there is no interruption, and on the third the whole
dorsal area is occupied by a large orange patch,
bisected for a part of its length by the deep black
dorsal line; and on the thirteenth the subdorsal
wedges are replaced by two large squarish marks;
the hairs are dark brown; the head a most brilliant
black.

Some of the larvæ had the orange marks very faint
indeed, and two of them had no orange marks at all
except on segments 2, 3, and 13, thus presenting a
good variety.

The pupa short, stout, reddish-brown in colour, the
anal segments still enveloped in the cast larva skin (I
notice this to be the case with the other species also),
enclosed in a thin web, in which bits of moss and
lichen were sometimes interwoven, and placed under
any protecting cover, such as a stone.

The moths I bred were very fine, much larger than
any I ever captured, and although varying somewhat
among themselves in the depth of their grey tints, yet
none of them were at all like *stramineola*. (J. H., 5,
9, 68; E.M.M. V, 110.)

LITHOSIA PYGMÆOLA.

Plate XLI, fig. 6.

On the 7th of June, 1862, Mr. Doubleday kindly sent me the larva of this species.

It was said to feed on lichens growing amongst moss, but lived only a few days, as I could not find any such lichen as the small portion of food which accompanied it, and which had a very pungent saline odour; it refused all other kinds of lichen and so starved.

It was short and rather thick, tapering a little posteriorly; brown on the back, with a thick black dorsal line; the subdorsal lines dark brown, and the sides rather paler brown, with a dirty-white line along the spiracles; the tubercles with short brown hairs, and the head black. (W. B.; E.M.M. I, 48.)

On the 10th of August, 1878, I received from Mr. W. H. Tugwell a good number of eggs of this species and a supply of the two species of lichen, on which the larvæ are known to feed. The eggs were all laid loose.

The shape of the egg is globular seen from above, but having a considerable depression beneath, apparently of a smooth surface, but really pitted so very minutely that even with a strong lens this character is scarcely appreciable. The colour is a very pale salmon or flesh-colour and very glossy. Without undergoing any further change they began to hatch on the evening of August 15th, and by the next morning about twenty were hatched; these and the remaining eggs were then placed on two pots of lichens.

The newly-hatched larva is very much the colour of the egg; it is rather stout, the head broad and large, brown on the crown, whitish in front above the mouth, which is brown; there are faint internal subdorsal lines of a brownish flesh-colour, extending as far behind as the tenth segment on which they are most distinct, and there unite across the back; a faintly darkish tinge of the same colour runs along the back from the head; the three hinder segments are almost colourless; the body is clothed with long whitish hairs.

When eight days old they had become of a light greenish-drab colour and seemed of a wandering disposition, as I found several on the outside of the pots.

After hybernation the larvæ were nowhere to be seen.

Another batch of eggs was received from Mr. Tugwell on the 21st August, 1879, just similar to the above; on the 30th they became more transparent and showed the embryo rather plainly through the shell as a dark grey spot. (W. B., Note Book III, 250.)

LITHOSIA MUSCERDA.

Plate XLI, fig. 7.

Eggs were received from Mr. Barrett on July 30th, 1870, and larvæ hatched on August 3rd. To these the same treatment and food was given as had already been tried with the larvæ of *Nudaria senex* (see p. 37). They hybernated small, when about one-fifth of an inch long; three were seen alive and feeding in February, 1871, these moulted at the end of March; two were then accidentally lost; the survivor moulted for the last time on May 6th, and was full-fed about the end of that month; it spun a cocoon, but had not strength to become a pupa.

Probably the right food for this larva is some sort of lichen growing on the sallow bushes in the soaking wet parts of the fens, where the moth occurs; Mr. Barrett noticed that it affects these bushes far more than any other kind of growth in the fens, and he observed that it is on the wing from early dusk till darkness sets in, when it disappears until midnight, after which hour it has another short flight; and probably there is a third flight in the morning dusk.

The egg was noted as small and shining. The young larva is of a dirty whitish colour, with black head, the tubercles furnished with single, stiff, dark hairs. When the larva is about one-fifth of an inch long the tubercles are shining black, and furnished with tufts of short hairs, the head shining black, the general colour of the body and hairs dull black, dorsal line and segmental folds velvety-black, a pair of dull orange spots on the second segment. This appearance continued up to the last moult; after that had taken place for an hour or two the colouring was very striking; the head was shining white, and while the tufts on the first segment and down the centre of the back were darkish brown, all the others were bright, light reddish-brown; but this gay dress was sobered down again.

The length of the full-grown larva is about three quarters of an inch, the figure rather stout, cylindrical, tapering only at the second segment and head, and again at the thirteenth; the legs well developed; eight tubercles on each segment raised and tufted, the front dorsal pair being only moderately large, but the hind pair much enlarged and transversely oval in shape; on segments 3 and 4 the front pair are larger than the hinder pair; all these tubercles thickly set with very short hairs. The general colouring is rusty black, the ground colour of the body being velvety blackish-brown, marbled with reddish-grey, the dorsal stripe and subdorsal line deep velvety-black; on each side of the dorsal line on the second segment, and again on the front of the thirteenth, is a squarish, dull, deep red spot; head shining black; tubercles and hairs all deep brown; each front pair of tubercles set in reddish-grey rings. There is a fine reddish-grey, interrupted subspiracular line; the belly pinkish-grey; all the legs shining, dark reddish-grey; tips of prolegs pellucid.

The larva retired into a curled-up bramble leaf, and
there formed a thin, webby cocoon of greyish silk,
outside which was a finer and thinner web of white
silk. (W. B. and J. H., 9, 12, 71; E.M.M. VIII,
173.)

A larva received from the Rev. J. Hellins in 1871
(date not given) was three-quarters of an inch long,
cylindrical, of tolerably uniform bulk, only the second
and thirteenth segments tapering; the head smaller
than the second segment; the legs and prolegs well
developed.

The warty tubercles in high relief in pairs, on each
segment down the back; the first pair moderately
small, the second pair very large and swollen, transversely oval; two rows of tubercles along the sides,
i. e. one in each row on each segment; across the
third and fourth segments the tubercles on the back
in pairs, the largest in front, and the smallest partly
behind it in an oblique direction.

The head is of a brilliant shining black; the body
a velvety blackish-brown, the dorsal and subdorsal
stripes deep black; the ground colour between them
is marbled with reddish-grey, and there is a fine, interrupted, reddish-grey, subspiracular line; the tubercles are deep brown, thickly beset with radiating
hairs of the same colour; the front pairs of small
tubercles are set in rings of reddish-grey. On the
second segment on each side of the dorsal stripe and
on the anterior portion of the thirteenth segment on
each side of the dorsal line, there is a squarish deep
dull red spot. The belly pinkish-grey. The legs and
prolegs dark reddish-grey and shining, the tips of the
latter pellucid.

This larva died after spinning its web, but another
in 1874 produced the moth June 24th. (W. B., Note
Book I, 78.)

ŒNISTIS QUADRA.

Plate XLII, fig. 1.

On the 30th of July, 1872, Mr. W. H. Harwood kindly sent me eggs of this species, laid close together on the side of a chip box; and he supplemented his gift by a few more, which came from a correspondent of his on the 8th of August, laid in clusters; in both instances some of the larvæ were hatched in transit, and all of them were out by the 13th.

The egg is hemispherical, most minutely pitted on its surface; of a rather glaucous bright green colour, turning olive, and again dark brown just before hatching; a large hole is eaten by the escaping larva in the upper part of the shell, which looks quite white when empty.

The young larva for a day or two is rather gelatinous looking, of a dirty whitish tint, but soon acquiring an internal pinkish tinge, showing a brown streak within the thoracic segments, the head being dark brown, and the body bearing some rather long, dirty whitish hairs. In about ten days the first moult takes place, when, as is the case with other species of *Lithosiæ*, so much of the characteristic marking and colouring of the mature larva is assumed, as suffices, even then, to distinguish it from its congeners; the whole larva now becomes tougher in texture, and the back becomes yellowish, prettily outlined with black, and with an interrupting spot on the eighth segment.

Unfortunately, I cannot give an account of the appearance during hibernation; both the young larvæ referred to above, and those also which on two other occasions I received from other friends, having died whilst no more than a quarter of an inch in length. However, I think that the smallest of the three I am now about to describe had, when first sent to me, scarcely increased in bulk since hibernation.

Mr. Harwood, still most kindly mindful to help me

with this species, sent me three young larvæ of varying
size, which he had beaten from oak trees near Colchester, and at St. Osyth, on the 10th, 16th, and 17th
of June, 1873.

These were kept separate, and their progress was as
follows :—No. 1, June 11th ; length three-quarters of
an inch, moulted 19th, increased to one and three-
eighths of an inch, spun up July 1st ; imago 21st, a
male. No. 2, June 18th ; length five-eighths of an
inch, moulted 21st, increased to three-quarters of an
inch, moulted 30th, increased to one and three-eighths
of an inch, spun up July 12th ; imago August 2nd,
a male. No. 3, June 18th ; length half an inch,
increased to five-eighths of an inch, moulted 24th, increased to three-quarters of an inch, moulted July 3rd,
increased to nearly one inch, moulted July 14th,
increased to one and a half inches or a little more,
spun up 27th ; imago August 14th, a female.

Each of these larvæ, on arrival, possessed all the
characters and colours that distinguished them through
their changes of skin to the adult state presently to be
described. The food supplied to them consisted of
various lichens from oak trees, and at first a few leaves
also, as I noticed the oak leaves that were sent to me
with each larva had been nibbled a little on the
journey; I also gave them *Lichen caninus*, for which
they soon showed such a decided preference that it
became almost their only nourishment; when disturbed,
they were very lively and active, running quickly over
any surface, yet clinging with a firm foot-hold when
they chose. Altogether, a great quantity of food was
devoured by them, and at times they seemed to eat
quite voraciously, always on the dark cuticle of the
lichen, not seeming to care for the pale fleshy substance beneath.

When about to moult, the colours became less vivid,
and the details less distinct ; at such times the larva
would leave its food for the leno cover of its cage, and
there spin a patch of silk, and fix itself upon it ; then

there seemed to ensue some operation of denuding itself of most of its hairs; but this process I was in every instance unable to witness, it being always effected during the night, generally the first night after the larva had taken its position on the silk; most of the hairs left remaining were on the second and third segments; nearly all the others appeared to have been bitten off close to the skin, excepting some few mere stumps of various lengths left along the sides. The actual moult would take place either on the first, second, or third night after this loss of hair, the minimum time with the smallest, the maximum with the largest larva; after moulting, the first meal was evidently made on the cast skin, as no trace of it could be found beyond the head piece, except in one instance, when a small fragment of skin remained. This breakfast on its old skin by a hairy larva was to me very surprising; it seemed, however, to act beneficially, for the next meal on lichen would be a hearty one.

I found that after each larva had attained its greatest length, it began gradually to shorten for three or four days before spinning its cocoon, although still occasionally feeding, sometimes even ravenously, during this period.

The full grown larva, as I have said, varied from one and three-eighths of an inch to nearly one and five-eighths of an inch in length; was moderately stout in proportion, somewhat cylindrical in figure, tapered a little from the fourth segment to the head, also from the eleventh to the anal extremity; the thoracic segments deeply wrinkled, the others plump and separated by well-defined divisions; the ventral prolegs long and well-developed, the anal prolegs long and extended behind beyond the end of the body; each segment with five prominent wart-like tubercles on either side, forming through the length of the body as many longitudinal rows; the two upper rows nearly close together along the subdorsal region, the others at equal distances along the sides, the lowest almost on the belly, all of them thickly furnished with long radiating hairs curved a little upwards at their tips.

The head is black and lustrous, the ground colour of the back a bright primrose yellow, which appears but little on the second segment, being there merely an edging and fine dorsal division to a blackish-grey mark ; this yellow is a little more seen on the third and fourth segments, where the large pairs of tubercles in front are black, the smaller ones behind them bright orange, the space between these on the fourth segment transversely barred with black, which more or less tinges the dorsal stripe, and produces a conspicuous central triangular or cruciform black spot ; the complex broad dorsal marking widens a little (diamond-like) on the middle of each following segment, and is composed of a fine broken grey outline, followed within by a line-like interval of the yellow ground, and then with freckles of bluish-grey edged with darker grey, and having a middle delicate thread-like interval of the yellow ; near the subdorsal region, run double fine broken lines of grey freckles, which on the front of the fourth segment are absent, but only interrupted on each of the other segments as they approach the bright orange tubercles placed in twos, i. e. a very small roundish one in front, and a large one transversely oval just behind it.*   Besides the thoracic black tubercles, mentioned, others occur on the eighth, twelfth, and thirteenth segments, as follows :—The small front tubercles on the eighth are black, and just there the dorsal region is also more or less black, together forming a conspicuous trilobed spot ; on the twelfth is a greater suffusion of the black, in which both large and small tubercles are dyed ; the tubercles on the front of the thirteenth segment are also black, the anal flap is dark brownish-grey, blotched with blackish and sparingly freckled with yellow ; the yellow ground of the back is very effectively relieved

* In the females these tubercles are *deep orange-red*, and the dorsal markings more decidedly of a diamond shape on each segment.—W. B.

by the broad subdorsal velvety-black stripe, on which the larger orange tubercles encroach. It has a very broken thread of yellow dots along the middle, and is margined below with a fine line of yellow, with another more interrupted beneath it; thence the ground colours of the side are dark reddish-grey, paler yellowish-grey nearer to the spiracular region, and darker brownish-grey below, including the semi-transparent ventral and anal prolegs with their brown hooks. The spiracular region is edged above and below at the segmental divisions with pale yellow; all the lateral tubercles are longitudinally oval and dark brownish-grey, each of the uppermost ones placed on a blackish crescentic blotch delicately edged with pale yellow; the belly dark greenish-grey, with a yellowish interrupted stripe on each side close to the prolegs; the hairs which hide the spiracles are chiefly grey, or slightly mixed with a few black ones on the sides, but those proceeding from the few dorsal black tubercles are blackish, and all are glossy. In one larva the lowest hairs along the sides were whity-brown, the next row above grey, and the upper rows darker grey mixed with black.

The pupal change, in one instance, occurred on the fourth day after the commencement of the cocoon, which was spun against the side of its cage, and in junction with the leno cover of it, and was formed of a large gossamer web of a roundish figure, about two by one and a half inches, of a darkish grey colour, and having the larval hairs interwoven; inside this outer web was a hammock of a finer-textured silk, held in suspension by fine threads at intervals in connection with the outer fabric. The pupa within the hammock lay belly upwards, and was eight lines in length, two and a half lines broad, almost uniform in size throughout, the head rounded, and only the last two segments tapered to the blunt and rounded tip; the surface smooth, quite black, and highly polished; the old larval skin lying detached behind it. (W. B., 7, 2, 74; E.M.M. X, 217.)

GNOPHRIA RUBRICOLLIS.

Plate XLII, fig. 2.

A tolerably abundant larva in beech woods during September and October, feeding on the tree lichens. I also found it once swarming on a lichen-covered park paling, and reared a large number of the perfect insects, which appeared during the month of May.

The larva is rather elongate, tapering posteriorly; head blackish, body greyish and freckled with yellow, a fine thread of whitish, bordered with grey, forms the dorsal line, which is white on the second segment, the subdorsal is a black line on the second, third, and fourth segments, and on the remainder becomes an elongated black trapezoidal mark on the anterior two-thirds of each, and terminates on the twelfth. The ground colour of the back on each side of the dorsal line of the seventh, eighth, ninth, and tenth segments is whitish; the sides mottled with greenish-yellow and grey; tubercles hairy. (W. B.; E.M.M. I, 49.)

CYBOSIA MESOMELLA.

Plate XLII, fig. 3.

On two or three previous occasions, I kept a larva or two alive from summer till after Christmas, having fed them on sallow leaves, green or decaying; and last spring I managed to retain one even until the new sallow leaves were out again, but it would not resume feeding after hibernation, and so died; it was then quite half an inch in length; in colour a velvety-black all over, and covered on every segment, save the head and second, with tufts of singular spatulate dark grey hairs. I should much like to procure some sort of food on which this species would feed up, for they would never take to any sort of lichen I gave them. (J. H., 5, 9, 68; E. M. M. V, 111.)

On May 5th and 25th, 1871, larvæ were received from Mr. Harwood, which had been found on the trunks of oak trees, feeding on a pale lichen growing intermixed with moss, but not sufficiently developed in its growth to enable us to make sure of its name. These larvæ soon spun up, and the moths, extremely fine examples, were bred on June 9th and 18th.

The full-grown larva is nearly an inch long, figure moderately stout, and tapering only at the head and second segment, and at the thirteenth segment. On each segment behind the second are eight raised tubercles densely tufted; the colour of the body is deep velvety slaty-blackish; the head shining black; a deep velvety-black patch on the second segment; the anterior legs shining black, the ventral prolegs pellucid, pale greyish, tipped with black; the second segment bears only simple black hairs, and similar hairs are found along the sides of the other segments just above the legs; but the tufts on their upper parts are composed of black hairs so densely feathered that they catch the light and receive quite a greyish effect from their peculiar softness, and almost entirely hide the skin beneath. In this peculiar featheriness of the larval clothing, this species comes so close to *Miltochrista miniata*, that it might well stand in the same genus with it; and it seems no improvement on the arrangement of Doubleday's List, in which they do actually stand close together, though in different genera, to separate them, as Staudinger has done, by the insertion of *irrorella* and others between them.

The stout pale brown pupa is enclosed in a comparatively large cocoon, formed of semi-transparent, thin, greyish silk web, spun in any convenient hollow under the moss or lichen. (W. B. and J. H., 9, 12, 71; E.M.M. VIII, 172.)

PHILEA IRRORELLA.

Plate XLII, fig. 4.

On July 30th, 1865, some eggs were received from Dr. Knaggs, and noted as globular, pearly in texture, clear purplish-brown in colour. The larvæ hatched August 13th, but no note of them was taken, and they must soon have perished for want of the proper food and treatment. There is no doubt, however, that in their natural habitat they must hybernate when small and feed up in the early summer.

On May 24th, 1867, after considerable search, a number were found, then approaching full growth, on the Sussex coast. The food is a blackish-brown lichen, growing on stones above high-water mark, and in some cases mixed with a yellow lichen, a fact of much interest when the colouring of the larva is considered. The larva seems fond of sunshine, moving about in it slowly over the stones; when about to moult it protects itself by spinning overhead a number of silken threads, under cover of which it remains until the moult is completed.

The moths were bred early in July.

When the larva is full grown its length is about six-eighths of an inch, the figure proportionate, moderately stout, tapering a little from the fourth segment to the head, and again at the thirteenth; six raised tubercles on each segment studded with longish hairs; the ground colour blackish-brown above, and dark reddish-grey or purplish-grey on the sides; belly and legs reddish. The dorsal stripe takes the form of a series of deep, brilliant yellow, acorn-shaped marks, the acorns pointing backwards, and so placed that the segmental folds mark the separation between the cup and the fruit; the paler and duller yellow subdorsal line much interrupted; the spiracular stripe of bright yellow also much interrupted; the raised tubercles blackish; the hairs blackish-brown; the ground on the back, and the lower part of the sides, minutely freckled with yellow; the inconspicuous spiracles dirty white, ringed with black.

The short, stoutish pupa, placed in a cocoon of thin webby silk, spun amongst the stones and *débris*. (W. B. and J. H., 9, 12, 71; E.M.M. VIII, 171.)

NUDARIA MUNDANA.

Plate XLIII, fig. 1.

The full-grown larva was received from Dr. White on the 31st of May, 1869, it having been captured feeding on lichens on an old stone wall.

Its length three-eighths of an inch, its figure rather stout in proportion, uniform in bulk throughout; the legs all well developed; six raised tubercles on each segment bearing long straggling fine hairs; the ground colour of the back bright sulphur-yellow; the dorsal stripe dark greyish-brown; a blackish dorsal spot on the eighth segment; the subdorsal line blackish-brown, the whole body below this, including the legs, of a semi-translucent, pale greyish-brown; all the tubercles and hairs rather dark greyish-brown; the head dark brown. (W. B. and J. H., 9, 12, 71; E.M.M. VIII, 171.)

NUDARIA SENEX.

Plate XLIII, fig. 2.

Eggs were received from Mr. Birks on July 18th, 1870; the larvæ hatched on the 21st; they fed on decayed sallow and bramble leaves, on the young growth of *Hypnum sericeum* and *Weissia cirrata*, and on *Lichen caninus*. The larvæ hybernated; the last moult took place early in May, 1871, and the larva was full-fed about the beginning of June; the imago appeared on the 23rd of June.

Mr. Birks described the locality in which the moths were captured by him as a swamp very rich in plants, and he found them either hovering over tufts of low herbage and coarse grass, or resting on the blades and stems of the grass or reeds. He could see no lichens, except on the trunks of the trees growing there, and he never noticed the moths haunting these, as we might suppose they sometimes would, if they deposited their eggs on them. Possibly the food may be some lichen growing under the herbage on the damp ground.

The female while laying her eggs mixes with them fluff from a tuft at her tail, which she detaches by means of her two hinder feet; and the way in which the fine plumes from this tuft adhere to the eggs makes it rather hard to describe them.

The larvæ, when hatched, were placed in a flower-pot with growing moss and lichen and straightway hid themselves; nothing more was seen of them till the solitary survivor of the whole brood was detected feeding early in May. Probably the rest were destroyed, while yet tender, by the small slugs and snails that infest lichens, and cannot be got rid of except by picking the latter to pieces; small centipedes also hide themselves away craftily, and no doubt do mischief.

The egg is small, globular in shape, but so soft that the outline is not at all regular, the shell shining, covered with faint irregular reticulations, yellowish in colour.

The young larva is pale grey, with a central olive stripe down the back, and with five or six long, pale grey hairs from each tubercle. Just before the last moult, the whole larva has a waxen, dull, smoky appearance; the tubercles raised and studded with tufts, formed of short smoky hairs, mixed with a few feathered plumes. When full grown the length is three-eighths of an inch, the figure very stout in proportion; the tufts so dense that the skin cannot be well seen, except when the larva curls itself up, and then it is seen at the segmental divisions—waxen-looking, and of a deep reddish-grey colour; the head shining black, the anterior legs glossy, tipped with black, the ventral prolegs translucent, in colour pale grey; the tuft-bearing tubercles are six in number on each segment; the tufts on the second segment are composed of single dark brown hairs, but the other tufts are much denser, and formed of two sorts of hairs, the more numerous being pale brown stiff hairs, with sharp black points, and being sparsely barbed or feathered; the others, fewer in number, are taller, with black stems, and densely feathered all round with soft, pale brown plumage.

The cocoon was of an oval form, about four lines in length, formed of close-spun silk and attached to the cover of the box in which the larva was confined; the hairs of the coat were all woven in, giving the cocoon a brown colour and rough texture.

The pupa skin, examined after the exit of the moth, was about one quarter of an inch in length, highly polished, of a rich deep brown colour, the segmental divisions showing as pale reddish rings. (W. B. and J. H., 9, 12, 71; E.M.M. VIII, 171.)

NOLA STRIGULA.

Plate XLIII, fig. 4.

I am indebted to the unvarying kindness of Mr. W. H. Harwood, of Colchester, for valuable information concerning the habits of this pretty species, and for opportunities of studying and describing its larva, examples of which I received from him on June 15th, 1869, and on June 8th, 1871.

These larvæ fed on oak, principally on the under cuticle of the leaves, and when full-fed, spun up in small boat-shaped cocoons of silk, about five-sixteenths of an inch in length, assimilating perfectly in colour to the surrounding surface of the bark on which they

were constructed,—a circumstance which rendered
their detection very difficult. After the escape of the
moths, which took place some time about the middle
of July, the cocoons still retained their form and
appearance.

For the sake of close examination, one individual
was kept without bark until too late for its spinning a
perfect cocoon, and at last it attached itself to the
underside of a leaf by the tail amongst a few threads,
and there pupated much after the manner of an
Ephyra.

The full-grown larva is but little more than three-
eighths of an inch in length; its body is rather stout
in proportion, thickest at the third and fourth segments, and tapered a little from the seventh to the
anal extremity; the head is full and rounded, but of
less bulk than the second segment; the body is
rounded on the back and sides, and rather flattened
beneath. It has three longitudinal rows of prominent
wart-like tubercles on each side, *i. e.* six on each segment, bearing fascicles of radiating hairs. It has
fourteen legs, the first ventral pair situated at the
eighth segment.

The colour of the body is pale buff, sometimes
partaking of a flesh tint; the dorsal stripe is yellowish
or whitish flesh colour, very broad, and well-defined
by a fine border line of brownish-grey; the subdorsal
line is brownish-grey, but interrupted at the segmental
divisions; all the tubercles are broadly ringed with
this colour. A conspicuous blackish-grey blotch covers
the back of the seventh segment and extends from
one subdorsal line to the other; there are indications
of other blotches of the same colour on the tenth and
eleventh segments, but these are cut in twain by the
broad, clear, pale dorsal stripe travelling through and
separating them into a narrow dark mark on each side
of the back. The sides are flesh-colour, the spiracles
are entirely hidden from observation by the numerous
hairs which diverge near them from the tubercles;

the ventral surface is pale flesh-colour and naked; the head is blackish-grey, the lobes narrowly margined in front with pale flesh-colour.

The hairs of the tubercles on the anterior segments are pale brown mixed with a few of dark grey, and some few of them in front of the second segment, and especially on the third, are very long; the tubercles on the rest of the body are furnished with hairs of a paler yellowish colour; on the back of the anterior part of the anal segment, issuing from each side, are a few hairs of extra length, which converge and taper on each side to a fine point directed outwards in a slightly downward curve, so that these two fine points of hair resemble a forked tail.

The pupa is four lines long, including the cast larva skin adhering to its tail; it is not very stout, of ordinary shape, though the wing-cases are long in proportion; these last are reddish-brown in colour, the other parts very dark brown and without much polish. W. B., E.M.M. IX, 15. (W. B., 5, 72; E.M.M. IX, 15.)

NOLA ALBULALIS.

Plate XLIII, fig. 5.

Six larvæ were sent to me by the Rev. J. Hellins, who had them from Mr. Platt Barrett; they arrived on June 20th. They were feeding on the leaves of the dewberry (*Rubus cæsius*), fearless little fellows, caring nothing for being tumbled about, so long as they were not deprived of the leaf on which they happened to be feeding.

The full-grown larvæ were over half an inch in length and stout in proportion; when stretched out in crawling they attained a length of five-eighths of an inch. They had six rows of round projecting tubercles, the lowest row just above the legs standing on thick basal stalks projected more than the others; each of these tubercles (six on a segment) bore little fascicles of hairs, the central hairs considerably longer than the others, that is, generally one or two on each tubercle excessively long; each segment was in itself plump and well defined, and with one transverse subdividing wrinkle very close to the segmental division.

In colour some were of an almost ivory whiteness, some of a pale flesh tint, others, again, of a bright pinkish-orange colour; in all there were twin dorsal light greyish lines and velvety black subdorsal marks, or blunt wedge-shaped spots in some commencing on the seventh segment and ending on the eleventh.

In one instance the lobes of the shining flesh-coloured head were marked with brown, and the plate on the second segment was brown, divided dorsally with the ground colour, the subdorsal black marks commencing faintly and small on segments 4 and 5, stronger on 5, and still stronger on 7, as a right-angled mark, thick at the angle, which extended across the back as far as the dorsal lines, the tubercles on this segment only being rose-pink. The right-angled black marks were continued, but rather thinner, on segments 8, 9, 10, and on 11, where they end; they are even thicker on the eleventh than on the seventh, thus forming nearly a bar of black across the beginning of the segment; the legs brown, the ventral prolegs tipped with brown, the fascicles of short hairs whitish, the longer hairs dark brown. The dorsal pair of tubercles on the eighth, ninth, and tenth segments finely outlined in part with black.

This last individual I watched at intervals constructing its curious cocoon, which is made on a dry grass stem of the thickness of a duck-quill; it commenced by gnawing in a line downwards at the longitudinal fibrous exterior of the stem, loosening it in a long thread, then another and another thread, and so on in succession until a great number of small threads were loosened of the length of nearly an inch, but not detached as yet. This gnawing was patiently and persistently continued round the stem, excepting only at that part which was intended to become enclosed. On this part the larva at length took up its position, and turning its head on one side, began to lay a glutinous secretion and detach and fasten upon it a short length of the previously loosened fibre, one length after another, longitudinally and parallel, and so continued to raise on either side of it structures much resembling a pair of wings (one on either side), concave within and tapered nearly to a point at the stem at each of their ends, the larva appearing to gum over the surface from which it removed the fibres.

On the night of June 29th I left it engaged in loosening some fibres near three-eighths of an inch above the top of the wing-like structures, its body behind resting within them; the next morning the larva had wholly retired within the two wings, and having drawn their edges together with silk, they united formed a somewhat fusiform cocoon much after the fashion of that of an *Anthrocera*. The larva when I saw it was engaged in filling up a few interstices at the junction with silk, some of the long dark hairs of the body being interwoven; but I could for some time plainly see through them the head of the larva at work, indeed, till 11 a.m., when they were all so thickly spun up as to prevent me from further watching, and only a slow and gentle intermittent throbbing of the cocoon proved that the occupant was still busy at work within; the assimilation to the stem itself in colour and texture was remarkable. The length of the cocoon was about seven-eighths of an inch; its upper surface rose tapering from the stem rather suddenly in a slope to the thickest part at about one-third of its length, where the head of the larva was last seen; thence it again tapered and sloped off very gradually at the lower end. (W. B., Note Book III, 96.)

NOLA CENTONALIS.

Plate XLIII, fig. 6.

On August 21st, 1879, I received twelve eggs of this species from Mr. Tugwell, who had been staying at Deal, where he found the species and saved two females for eggs.

The food-plant of the larva not then known, but the principal plants the moths were found *on* or near were dwarf plants of *Hippophaë rhamnoides*, *Salix argentea*, *Senecio jacobœa*, *Pimpinella saxifraga*, *Lotus corniculatus*, *Thymus serpyllum*, and of grasses *Arundo arenaria* (Marram), *Agrostis* —?, *Aira* —?, *Triticum* —?, and one or two *Juncaceœ*. Once a moth was taken by Mr. Tugwell on Freshwater Down, where *Thymus serpyllum* grows very freely.

The eggs were laid on and adhering to a paper box singly and in groups of two or three or four in a cluster. In shape the egg is circular, somewhat flattened in character, and having a depression above. It is very strongly ribbed and most minutely reticulated; the shell is glistening and of a pale, creamy-white colour. Unfortunately these eggs proved infertile.

Mr. Tugwell informed me that the fertile eggs showed on the fourth or fifth day a central pale grey or round spot, which increased in intensity till the hatching, and the other parts of the egg became less white.

On September 16th Mr. Tugwell sent me six young larvæ feeding on *Trifolium procumbens* and *Lotus corniculatus*; they were little more than a tenth of an inch in length, rather stout in proportion, deep flesh-colour, the head shining reddish-brown, marked on each lobe with darker brown, a darkish brown plate on the second segment; the body with subdorsal, lateral, and spiracular rows of projecting warty pink tubercles, with radiating, longish greyish-brown hairs.

This larva has only three pairs of ventral prolegs, the first pair being absent.

Trifolium procumbens appears to be their proper food, as they like getting on the hop-like seeding heads, to which their colour assimilates well.

On the 22nd and 23rd they moulted, and the next day were thicker and of a subdued, velvety, deep red colour, the tubercles a glistening blackish-brown, the hairs light brown. They now ate away the cuticle of the leaves, causing semi-transparent whitish blotches on them.

(From Mr. Tugwell I learnt subsequently that the eggs hatched August 27th and 28th, and the larvæ moulted first on September 5th and 6th; the second moult occurred from the 13th to the 16th, and contemporary with my portion of them the third moult occurred from the 21st to the 25th.)

On the 26th the larvæ had again become paler flesh colour, but with somewhat of a faint greenish tinge, the tubercles dark reddish. I observed one feeding on a blossom and another eating the red-brown, hop-like envelopes, exposing the ends of the seeds.

On September 30th one fixed itself as if for the fourth moult, and by October 6th all five had moulted, and, as on each former moult, the head-piece remained attached to the old skin. As this breaks away behind the plate, the larva draws its head out from the old helmet and then creeps out of its old coat; the split is somewhat in form of the letter I, which opens, and a portion lies over on either side, so that the egress of the larva is comparatively easy.

The larvæ were now feeding slowly on the flowers and leaves of *Trifolium minus*. They were dingy red, with a dorsal line of paler brownish-ochreous faintly edged with darker than the ground, the tubercles all of a shining blackish-brown, the skin of the body soft without gloss; they were again given *T. procumbens*,

of which they ate well both flowers and leaves, until the
15th, when, a frost occurring, they became very torpid.

Their structure was now very well seen, they having
only three pair of ventral prolegs ; the head blackish-
brown and shining, a small semilunar plate of the
same colour and finely divided on the second segment,
and three rows of dark brown tubercles on each side
of the body ; the upper or subdorsal row was the
largest and most prominent, and all were thickly
studded with radiating brown hairs, a few single, much
longer, hairs occurring behind and along the sides.

On October 22rd, as they had not fed for more than
a week, I placed them out on a potted plant of *Lotus
corniculatus*, but next morning, the 24th, about 11
o'clock, one of two kept back in a bottle for observation moulted for the fifth time, whereupon I withdrew
the sprays on which were the other three from the
Lotus, and secured them in another bottle to await
their moult, which occurred during the night, so that
on the following morning, the 25th, I found that only
one larva was waiting for this operation, which was
not accomplished until 10 a.m. on the 27th. Up to
that date none of the larvæ had fed at all since
moulting, and up to the end of the month they all
refused to feed, and had begun to hibernate.

On November 4th I received a full-grown larva of
the same brood from Mr. Tugwell, who by keeping
several larvæ in a warm room with a fire and gas,
confined in a wide-mouthed glass bottle containing
mixed food of *Trifolium minus*, *T. pratense* blossoms,
and *Medicago lupulina*, on all of which they fed (the
bottle tied over with calico and over the bottle a large
glass globe (? shade), to prevent the food drying too
much), had forced them forward to maturity. This
larva I figured as soon as it arrived.

Its length was nine-sixteenths of an inch, or nearly
five-eighths when stretched out fully ; it was stout
and plump, the segments well divided, and having on
each side of the body three rows of prominent warty

NOLA CENTONALIS.   47

tubercles (*i. e.* a transverse series of six across each
segment), the upper row the largest, the middle row
the smallest, and all emitting fascicles of radiating
hairs, single hairs of much greater length proceeding
from the anal tubercles, and one from each tubercle of
the lowermost row on the sides. The skin itself was
soft and smooth between these tubercles, which allowed
but little of it to be seen on the upper parts of the
body, but the ventral surface showed it to be soft and
just a trifle glistening; the body tapered a little
behind, and rather more from second segment to the
head.

It had only three pairs of ventral prolegs.

The colour of the glossy head was blackish-brown,
above the mouth a broad streak of dingy pink, papillæ
paler; the body generally and the tubercles were of
this same dingy deep pink or dirty purplish pink,
beautifully relieved by a dorsal line of ochreous yellow,
which passed through a V mark of velvety-black at
the beginning of each segment, separating the black
into two sharp wedges. A fine linear black streak anteriorly ran beneath each upper tubercle, representative
of the subdorsal line; the tracheal thread of faint pale
yellowish ran midway between the middle and lower
rows of tubercles, and showed more distinctly at each
part bearing a black spiracle directly on it, so that
this very faint yellowish line had an interrupted
appearance towards each segmental division. The
hairs were lightish greyish-brown, rather shining, and
from their reflecting the colour of the body and
of themselves also, and occupying so much space
about the upper surface of the body, produced
there somewhat of a general brownish effect. The
anterior legs dingy pink, marked with darkish
brown; the ventral and anal prolegs of the pinkish
ground colour, but more transparent and fringed
with hooks.

After feeding on flowers of common red clover for
several days it at length spun itself up after very

leisurely constructing the case formed of gnawings of
the bark of the stem in two wing-like halves, and
then drawing them together, the stem itself being the
base of the inverted boat-shaped structure.

As to my small hibernating larvæ I found one on
March 7th, 1880, but it went asleep next day without
feeding, but a few days later while in a bottle it began
to feed on young leaves of *Trifolium procumbens* for a
couple of days, and then ceased during an interval of
cold east winds, and remained fixed on the under side
of a leaf, where on April 1st it had moulted, the old
skin remaining close beside it.   This to my astonish-
ment made the sixth moult, a circumstance unique
in my experience!

From a communication since received from Mr.
Tugwell I believe it might be possible that the larvæ
sent to me were not in their second skins as he had
thought at the time, but really *before their second*
moult.

The larva then fed very sparingly, but on April
17th, though it had grown very slowly in the interval,
its colouring was different, and it agreed more closely
with the forced full-grown larva in having the yellow
ochreous dorsal line visible ; the general colouring of
the body was a darker red than before, the tubercles
were still dark blackish-brown, but the dark brown
hairs were more conspicuous.

On the 22nd it fixed itself for another moult (the
seventh, H. T. S.), which was accomplished on April
27th at midnight.

After a few days it began to feed again, and so continued at intervals according to the temperature until
May 13th ; the black V's at the beginning of each
segment through which the yellow dorsal line ran
had now become deep pink, and a ring of pale flesh
colour surrounded the base of every tubercle, all of
which remained blackish-brown.   It now ceased to
feed, and thinking it might spin up I supplied it with
a stem, which, proving to be hollow, the larva crept

inside, and there remained until the morning of the 18th, when it was out and feeding on a flower of purple clover. On splitting up the stem I found therein the cast-off skin (the eighth, H. T. S.).

On the 23rd it again fixed itself to moult (the ninth, H. T. S.), and accomplished the operation by the morning of the 27th. The black V's were now distinct, the tubercles no longer dark, the yellow ochre dorsal line quite bright. It fed well on flowers of *Trifolium pratense* and *repens* until June 3rd, when it ceased to feed, and on the 4th took up its position on a dry stem placed for it, and began very deliberately to construct its case of two side wing-like pieces of silk, covering them with portions gnawed off from the stem; these by the evening of the 5th seemed to be complete, and by night the larva was joining them together from within. By the next morning, June 6th, the larva was enclosed, and the structure appeared complete. (W. B., Note Book III, 296.)

[Surprise is expressed at this larva moulting the *sixth* time, but no surprise seems to have been caused by its moulting three times after the sixth moult!— H. T. S.]

On October 29th, 1882, Mr. W. R. Jeffrey sent me a few eggs laid by a bred female, which had paired in captivity. They were laid on a dry stalk by the side of a cocoon from which one of the moths had emerged, and adhered to the side of a little hollow channel so as not readily to be seen.

The shape of the egg is round and slightly flattened, having a central depression on the upper surface. It is finely ribbed and reticulated, white when first laid, but afterwards became of a more creamy tint, and having within the central hollow in the top of the egg a ring of brown atoms.

On November 4th I described a larva of this species, then nine-sixteenths of an inch in length stoutish, plump, with three rows of prominent tubercular warts, the upper or subdorsal row the largest, the next row the smallest, under them are the black spiracles, then another row of tubercles; the head and second segment tapering, the head smallest, dark, or blackish-brown, very glossy; colour of the body a kid-leather-like skin, a dingy pink, or dingy purplish-pink; the straight dorsal line ochreous-yellow, with a black velvety V-mark at the beginning of each segment through which the dorsal ochreous yellow line runs, separating each black V into two wedges. A black streak or line bounded anteriorly the lower margin of each subdorsal tubercle, and another, shorter and less noticeable, bounded the subspiracular tubercular. The tubercles were so large in proportion, and occupied so much space, that but little of the soft skin was seen between them, and a general brownish effect was produced by the rather brownish hairs which radiate from them. As usual in this genus the longer hairs were single, issuing from the lower tubercles, and noticeably from those of the thirteenth segment. A faint appearance of a paler ochreous line along the spiracles was most noticeable where the spiracles occur on it, so that this faint yellow line had an interrupted appearance. (W. B., Note Book IV, 175.)

SPILOSOMA LUBRICEPEDA.

Plate XLV, fig. 4.

On the 3rd of July, 1879, I chanced to find sixty eggs laid on the underside of a hawthorn leaf, side by side close together in a group; they were globular and apparently smooth-shelled, glossy, and of an opaque whitish colour; by their being of a good size I had the wish to prove their identity.

On the 12th they seemed rather less white and less glossy, and on the morning of the 14th they had turned to a light grey colour, with a dark grey blotch or two and with fine, black, hair-like equa-

torial lines showing through the shells. By 3 p.m.
the same day fourteen were hatched and most of the
other shells chipped.

The newly-hatched larva was very pale, just lightly
tinged with greenish-drab, and with a shining brown
head, and with rather long blackish hairs on the body.
After a moult the larva became rather more tinged
with warmish yellow-drab, and showed an internal greyish spot towards the hinder two or three segments.
After another moult it continued much of the same
colour. After the third moult it was half an inch long,
and the back was broadly marked its full width with
deeper drab, less transparent than the rest of the body,
and edged on either side with darker drab, and this
again below with whitish; there was a dorsal line of
whitish; the rest of the body and head as before;
tubercular dots whitish, hairs dusky.

Directly after the fourth moult (August 19th) it was
five-eighths of an inch long, and stout in proportion;
head yellowish-drab and shining, second segment
similar with the front margin paler; the rest of the
body greyish-drab, darker on the back, and broadly
edged with dark grey drab on either side, and again
close beneath with greyish-white in a ragged manner
of long, wedge-shaped marks united, the narrow apex
of one joining to the broad base at the inner edge of
the other on the segment before it; the raised tubercles
round and warty, bearing fascicles of radiating hairs
of black and of greyish; a dorsal line of greyish-white
or of whitish-greyish-yellowish, was outlined or edged
with darkish drab; spiracles whitish.

After the fifth moult on August 27th, it was in its
adult dress of dark olive-greyish-brown on the back
without gloss, and edged with darker brown next the
cream-white spiracular stripe, the whitish spiracles
close beneath it; the belly similar in colour and of
kid-glove texture as the back, but rather less dark;
the wart-like, round tubercles the same, all thickly
studded with radiating darkish brown glossy hairs.

The head was much as before, greyish-yellow and glossy, the mouth blackish, and a creamy-whitish streak above it; papillæ pale cream colour, ocelli black.

In many there was no trace of any dorsal line, while in some there was a very faint, paler, and interrupted line, visible chiefly on some of the hinder segments, and in others a cream-coloured line edged softly with dark grey was distinct on the third, fourth, fifth, and sixth, and less and less visible on the remaining segments; a trace of it there could only be detected by aid of a lens.

On the 8th of September most of them were spun up for pupation in cocoons of silk, in which the hairs from their bodies were interwoven. (W. B., Note Book III, 276.)

DEIOPEIA PULCHELLA.

Plate XLVI, fig. 3.

On the 18th of June, 1878, I received from Mr. W. H. Tugwell two larvæ in their third moult, which he had reared thus far on *Myosotis palustris* with others from a batch of eggs sent him from Mentone by Mr. J. Sidebotham. The eggs had reached him on the 24th May, and hatched the same day.

The two larvæ I received on the 18th of June were then respectively of the length of half an inch, and three-quarters of an inch, the segments plump and well cut at the divisions, the head the smallest. The ground colour was brownish-black or very dark brownish-grey, appearing black by force of contrast with an ornamental white dorsal stripe, which on the smaller larva was continuous, but not quite so in the other; in both specimens it was interrupted by a transverse bar of lurid red, or dingy ochreous-red, narrow in the larger one; there was a blotch of the same red below on the side; the spiracular region was puckered and marked

with a marbling of white streaks in a linear direction
along its length, the belly and legs of the same dark
colour as the ground of the back; the head a lurid
dark red marked broadly with black on the front of
each lobe. The tubercular spots were prominent deep
black warts, each bearing a curved, black, bristly hair,
those on the sides above the legs and some few above
on the thoracic segments were rather longer, whitish,
and finely plumose throughout their length; the whole
surface of the body shining.

Mr. Tugwell describes their first juvenile skin as
dull orange, having a few bristly hairs, and their
appearance maggoty and manner sluggish; their
second skin as darker greyish, with a transverse band
of dull orange on each segment from one spiracle
across the back to the other, the usual tubercular
warts black in each band and with a bristly hair. He
observed that they ate the leaves of *Myosotis palustris*,
but seemed to prefer the flowers and young seeds.

My two larvæ moulted on the 21st and 24th of
June. I figured the former on the 25th; they had
both ceased to show any red markings The largest at
this date measured seven-eighths of an inch and was
velvety black, softening into a lighter blackish-blue at
the segmental divisions, which added to the velvety
appearance; the dorsal ivory-white marking formed a
series of ornamental spots upon each segmental division; there was a small white speck on either side of
the front of a segment. Along the spiracular region
ran a ragged and rather interrupted white line, branch-
ing a little lichen-like upwards at each segmental
division, where it melted into the bluish-black; the
spiracles were black; the belly black and velvety, the
ventral and anal prolegs dark brownish-grey. The
black hairs of the back and the white secondary dorsal
hairs of the thoracic segments and those all along the
sides were long, pointed, and barbed or plumose as
before; the head was reddish, marked with black on
the crown of each lobe, the triangular space between

them finely outlined with white; the anterior legs black and shining.

The full-fed larva was about one inch long; the grey of the segmental divisions spreading a little very faintly over the other parts, which then showed large velvety black blotchy spots on the middle of the back of each segment and at the sides, and black warts rather shining at the base of the hairs, and each pair of hinder dorsal of these bordered in front by a thin margin of dark, rather rusty, or brick-red, thicker behind each of the side tubercles (these were the remains of the previous red band across each segment, showing narrow between the trapezoidals, and wider behind the side wart); along the belly was a broad central stripe of dirty pinkish, having a few black dots on those segments which bear no legs; on each segment near the beginning of the white dorsal marking were two isolated white specks, one on either side, sometimes just connected with the dorsal mark by the finest possible thread of white; and the same occurred at the hinder part of the segment behind the posterior pair of the trapezoidal tubercles.

The earliest of the two was full-fed by June 29th, and the other by July.* The former spun a few threads, drawing two leaves partly together with very open meshes, and pupated therein. This pupa was barely half an inch long, and one-eighth of an inch in diameter at the thickest part across the ends of the wing covers, from which point the three flexible rings were deeply cut, but with very slight tapering of the form of the abdomen, which at the end was very bluntly rounded off and furnished with eight fine-pointed bristles; the head tolerably well defined from the thorax, and this again also from the back of the abdomen, which swelled out thence. I figured the pupa on the 5th of July; the colour was then a very rich orange-brown, the eye-pieces black, wing-covers margined with darker brown, and the nervures strongly

* A blank left for the date.—H. T. S.

marked with blackish-brown, this latter colour occupying the whole of the tips of the wings; on the thorax
was a broad, dumpy, U-shaped blackish-brown mark,
and a narrow transverse linear bar of the same blackish-
brown across the middle of each segment on the back
of the abdomen, and a darkish brown line and similar
transverse bars on the ventral surface; the whole pupa
very glossy.

On the 25th of July, 1878, I received from Mr.
Tugwell fourteen eggs, part of a batch of a hundred
laid by a bred female, which had paired; they were
laid singly on coarse muslin. The shape of the egg
is globular, but depressed a little beneath; it is apparently smooth-surfaced. On their arrival they were
yellow, some a deep yellow, almost orange, and very
shining. The next day seven had changed colour to
darkish grey-brown, showing a blackish-grey blotch
in the centre beneath the top of the egg; three of the
others showed central faint orange-brownish similar
blotches through their yellow shells, and two were
shrivelling; in the evening of the same day (July 26th)
seven larvæ were hatched.

When first hatched the larva was of a dark grey-brown
colour, with a faintly paler dorsal line; head blackish,
and a blackish plate on the second segment; minute
black dots and hairs on the body. The first moult
occurred on July 31st and August 1st, when the larvæ
became orange-ochreous in colour, with minute black
dots as before; the second moult occurred August 3rd,
4th, and 5th, for some were already in advance of the
others. The larva was now brownish-orange, of rather
a dingy hue, with a faint pale dorsal line, and a broader
faint pale spiracular stripe; the head, the small plate
on the second segment, and the tubercular warts
blackish-brown. The third moult was attained by
the most forward larva August 5th, another was waiting for the operation, the other four not being so far
advanced, indeed, one had a few hours ago completed
its second moult. I saw one larva eating part of its

cast skin. After the third moult the colour changed
to a dark lurid purple, with faint whitish dorsal and
spiracular stripes, the head, small plate on second
segment, and tubercular warts blackish as before.
At this time it fed on the leaves and gnawed the rind
of the stalks.

A lens now showed the hairs to be barbed, those on
the back black, and the lateral ones white, assimilating well with the hairs of the plant, and as parts of
some of the leaves turn blackish the larva is protected
by its resemblance in colour.

As the larva grows in its third coat the brick-red
transverse bands are developed across the middle of
the segments. On the 21st of August the most forward larva was laid up, waiting for its fourth moult,
which occurred early in the morning of the 23rd; the
larva then entered the black stage, with scarcely a
trace of the red bands; the second larva had also
developed its third coat into an orange-buff colour,
with white dorsal and spiracular stripes and black
tubercular spots. It moulted August 26th, and was
then blacker at the spots than before, with the white
occupying more space at each end of a segment, where
a little smoky-grey appears at the sides, thus restrict-
ing the orange-buff to a transverse band across the
back as far as the spiracular region. This was now
over its fourth moult, consequently in its fifth coat.
(W. B., Note Book III, 236.)

LASIOCAMPA QUERCUS.

Plate XLVII, fig. 2.

On the 8th of June, 1868, Mr. Doubleday sent me
two larvæ found at Epping, which he assured me
were the true *L. quercus.* They came to me with
hawthorn.

Without going into a lengthened description of

structure and disposition of hairs, it will suffice to mention the characters which chiefly distinguished these larvæ.

The general colour of the hairs above the subdorsal stripe was of a rather light brown, but below and on the ventral surface darker brown. Along the subdorsal region there was a large triangular mark or streak of white with black centre. Besides the much-interrupted white subdorsal streak on the third and fourth segments, there was just above it on the black rings a whitish oval or shuttle-shaped mark with a black spot within it near its anterior margin. The white subdorsal stripe appeared to be *continuous* on all the black velvety parts of the body that were not hidden with hairs, for these interrupted it from view about the middle of each segment; from the subdorsal stripe oblique white streaks flowed backwards near the beginning of each segment; those beyond the thoracic ones were spotted or mottled with dull red below. *Above the legs were two whitish and red dull stripes, with indications of an interrupted middle line between them.* The oval spiracles were white. (W. B., Note Book II, 185.)

On the 25th of August, 1875, a pair of this species *in cop.* were brought to me, and the female laid a great number of eggs all loose in a box.

The egg is large, of a regular oval form, smooth and shining, of a pale drab colour, irregularly blotched with light brown. A few days before hatching they become wholly brown, and after the extrusion of the larva the shell regains its previous colouring.

The larvæ were hatched on the 12th of September, and were a quarter of an inch long, and hairy. (W. B., Note Book III, 37.)

LASIOCAMPA QUERCUS, var. CALLUNÆ.

Plate XLVII, fig. 3.

On the 8th of June, 1868, seven of these larvæ were sent me by Mr. Doubleday; they were feeding on hawthorn.

The general colour of the hairs above the subdorsal region was a bright golden brown, darker or lighter in individuals, but very bright and glossy; below, on the sides, the hairs were deep blackish-brown, the ventral surface still darker brown and nearly black. The triangular subdorsal mark on the second segment was just edged above with white, but was chiefly bright red, and the same with the ear-like subdorsal marks, of which one was about the middle of the third and one on the fourth segment.

There was no subdorsal stripe, but only a row of subdorsal spots, one behind each segmental division; these spots had much the character of a triangular oblique streak of red, having in some individuals the anterior apex white, but in others wholly bright red. There was *only one stripe above the legs;* this was red and only obscurely visible. The oval spiracles white, with a fine hair-like black line down the centre, marking the aperture. (W. B., Note Book II, 186.)

PŒCILOCAMPA POPULI.

Plate XLVIII, fig. 2.

On the 11th of June, 1874, I found a full-grown larva at rest on a branch of birch at Emsworth; it was about an inch and three-quarters in length, rounded above, rather flattened beneath; the ventral prolegs rather sprawling; the head full and rounded, but decidedly smaller than the second segment; the body of uniform size or nearly so, though when stretched out and walking it tapered a very little in front and rather more at the two hinder segments.

The head was bluish-grey, freckled with reddish- and brownish-grey; the front of the second segment margined with bluish-grey, followed by a fusiform mark of brown divided dorsally by a pale line. On the back of the other segments was a series of dark grey blotches, bearing the forms of inverted urns; they were freckled with blackish atoms; their apices placed hindmost were the darkest. Through these marks ran the darker dorsal line, and within them, on either side of the dorsal line, were two acute angular marks of bright ochreous-orange extending trans- versely; the sides below were somewhat crenulated with curves of dark grey on a whitish ground, and above them was a large blotch or suffusion of ochreous- orange, freckled with dark grey and surrounded above with a blotch in front and a larger blotch behind, of squarish form, dark grey, finely freckled with blackish like the curved blotches below. The oval spiracles were blackish, with a narrow whitish centre but very inconspicuous; the ventral prolegs were pale grey, streaked with darker grey, and tipped with brownish- grey hooks.

The dorsal marks on the third and fourth segments were blackish and rather conspicuously relieved by a marginal side blotch behind of whitish, which was but faintly indicated on the other segments. On the third and fourth segments were oblique side streaks of dark grey downwards and forwards. At the end of the sixth and beginning of the seventh segments the dark dorsal blotches were relieved on either side by con- spicuous large whitish blotches. The belly and inner surface of the ventral prolegs were buff-yellow, and each segment had a central black spot (much larger on the leg-bearing segments) in the middle of the body; on either side of the front of the second seg- ment was a round wart-like tubercle. The whole upper surface of the body and head was covered with a fine pubescence, the sides being fringed below with more numerous and rather longer grey hairs, interspersed with a few still longer of a dark brown colour. The surface of the belly very slightly pubescent.

On the 14th of June it was shortened and had begun to spin the threads for its cocoon. (W. B., Note Book II, 72.)

ODONESTIS POTATORIA.

Plate L, fig. 3.

On the 6th of August, 1876, a female moth was taken in the dusk of evening by Mr. Henry Terry in his hand as she flew past the door. He brought her into me to see what his capture was, and found she had laid twelve eggs whilst in his hand; these I kept for description.

The egg is rather large, roundish-ovate, some being of a rounder shape than others, having a small, very shallow circular depression above, and a deeper one beneath, the surface smooth, the colour opaque shining white, like porcelain; the depressions above and below are light greyish-green, and also the zones which at some distance surround each depression, and a smallish depressed round spot of the same colour is at one end of the egg, midway between the two zones. (W. B., Note Book III, 141.)

ENDROMIS VERSICOLOR.

Plate LI, fig. 3.

A long-cherished desire of obtaining eggs of this species, for the purpose of watching the larva through all its stages, was gratified on the 6th of May, 1881, when a dozen, laid on bits of paper and birch twig, were sent me by Mr. H. McArthur from Rannoch.

The larvæ began to hatch in the early morning of May 22nd, and continued to appear at intervals throughout that day and up to the next morning, when the two latest were hatched.

At once the young larvæ took readily to birch as their food, and moulted the first time on the 28th to 30th of the month; on the 3rd of June most of them had again moulted, and on the 9th and 10th they moulted the third time; their last moult (the fourth) began on the 17th, and concluded within a few following days.

Full growth was attained by some on the 26th of June, and from this date onward the remainder matured at intervals one after another until the 9th of July, when the last larva retired into the moss provided for the purpose.

In 1882 I was prevented from looking into their cage until the 1st of April, when I saw some specimens had already been out some time, as three or four were dead and much shattered ; after this, on the 3rd, a male and two females emerged, and another female on the 8th; three pupæ remained over until the present year, 1883, when on April 12th a male was bred, followed on the 18th by another, and on the 21st by a female, the males being much finer specimens than those of the previous year.

The egg of *versicolor* is of a good size, about 2 mm. in length, and rather more than 1 mm. wide, in shape much like that of a brick with rounded-off angles, slightly depressed on the upper side, sometimes on both sides, the surface apparently smooth and very glossy ; when first laid, it is of a light green colour, but this, in the course of a few days, changes to a dark brownish-purple, much the colour of a fresh birch twig. This lasts for about fifteen days and it then assumes a purplish-violet tint, gleaming like an amethyst, and the interior sems a little cloudy ; a few hours later, it is fainter and pinkish, and then the larva soon hatches. The empty shell, with the circular hole of egress at one end, still retains a faint tinge of pinkish-violet after the larva has escaped.

On leaving the egg-shell, the larva is a stout and robust creature of cylindrical figure ; the head, as usual at this time, the largest segment, is of a dark black colour, with greenish mouth ; the body velvety black, with a dingy olive-greenish plate on the second segment, having a wide black dorsal division ; on the other segments are olivaceous greenish-yellow, tubercular warts, each anterior pair on the back being distinctly larger than the others, which are very minute, all bearing a few weak, soft yellowish hairs. A black dorsal blunt projection is on the twelfth segment; the anal plate and outer sides of the anal legs are pale olive-greenish-yellow ; the ventral prolegs are blackish on the outside with greenish inner side, the anterior legs olivaceous yellow and shining.

From the first, they at intervals fed on two particular leaves near the top of the birch spray, whereon they had all assembled, holding to the twig by their ventral anal prolegs only, the fore part of each body being bent back away from the twig, leaving the anterior legs free ; by the fourth day their colouring had become dingy blackish-olive, with the mouth orange-ochreous, a blackish dorsal line, black tubercular spots, a conical hump on the twelfth segment, a faintly paler spiracular ridge on the thoracic region, and the anterior legs pale orange, with black bases.

After the first moult, the ground colour was of a subdued green, thickly freckled with black atoms ; the head and plate on second segment paler, of sober greenish-yellow, as were also the spiracular ridge on the thoracic segments, and a green backward-slanting stripe on the side of each of the others, and this was still paler and yellower on the eleventh and twelfth, on which last a stripe began at the top of the blunt eminence ; the anal flap was margined with the same colour ; the head was marked with two black stripes on either side ; a black dorsal line divided the front plate and continued throughout over the hump as far as the anal flap.

After the second moult, they were an inch long, and then broke up their society, and separated for independent existence, yet were sufficiently amiable, whenever they chanced to find themselves near each other, to agree perfectly well at any time. While resting, they still elevated the front part of their bodies as when younger. At this stage, the colour of the back was much lighter green, the dorsal line dark green, except at the apex of the hump, where it was black ; the sides were of a fuller green finely dotted with black. On the back the dots showed greenish, though they had become nearly obsolete there ; the stripes on the head were alternately whitish-yellow and dark green, on the thoracic segments the whitish spiracular ridge was conspicuous, as also on the other segments were the side stripes of yellowish-white bordered above with deep green, and these also now not only reached the segmental division in their downward slant, but crossed it, and were thence continued narrowly and obscurely below on the segment following.

After the third moult, their growth was quick ; two days' feeding increased the length from 1 inch 3 lines to 1 inch 4$\frac{1}{2}$ lines, with greater stoutness also in proportion, the thoracic segments decidedly tapering to the small head ; the relative colouring much as before, paler whitish-yellow-green on the back, with deeper green dorsal line, black at top of the prolonged hump, which was now seen to be slightly divided into two blunt points ; the yellowish side stripes margined both above and below with deep green, and the sides below them of still deeper green, irrorated with fine black dots, except just where the attenuated continuations of the side stripes could be traced; the bases of the anterior legs black.

After the fourth and last moult, their docile behaviour continued to be remarkable, as they showed no disinclination to be handled, but grew quite lethargic, often sleeping side by side contentedly like so many fat pigs ; but when awake they made good use of their time, consuming a great quantity of birch, and their growth was commensurate, for by the 26th of June some were 2 inches 3 lines in length, others, later, as much as 2 inches 7 lines, and bulky in proportion ; the head very small, with the thoracic segments rapidly tapering to it, and retractile as in *Chœrocampa*, though to a less extent.

The middle of the body is rather the thickest, and the twelfth segment, with its humped elevation, bluntly pointed and slightly divided, slopes backward at an angle to the anal flap ; the ventral and anal prolegs are developed much after the fashion of *Smerinthus ;* the other segments are lightly subdivided into four nearly equal portions by slight wrinkles, the segmental divisions more strongly defined, especially on the belly. The skin is soft and smooth, glistening on the head, which is green, and has two whitish or yellowish-white stripes beginning on either side, and continuing to the end of the thoracic segments, the uppermost as a sub-dorsal, and the lower as an inflated spiracular stripe ; the back is pale opaque green, slightly inclining to yellowish in the lightest, and to bluish in the deepest portions and in the dorsal line; below the yellow stripes, which are bordered above with green, the ground colour of the sides is of a very deep and rich full green, increased in depth by the close irroration of minute black dots, and relieved by the white oval spiracles delicately outlined with black ; in front of these comes a thin line of quiet ochreous-greenish, as though a continuation of the slanting stripe from the preceding segment, more noticeable on approaching the ventral prolegs, which, like the base of the anal pair, are bright crimson ; the whitish-yellow stripe on the side of the eleventh segment continues downward beneath the spiracle on the twelfth. From the top of the white horn-like hump, which is divided by a thin line of black, a whitish stripe descends on either side in

a slight backward curve, and the anal flap is margined
with yellowish ; the anterior legs are pale green, sometimes tipped with red, and with a black hook.

When full-fed, all the green colours of the larva
change to brown, and it becomes restless until it finds
the moss and leaves needful for its retirement and the
construction of its cocoon. The cocoon varies in
length from 1 inch 4 lines to 1 inch 7 lines, and is of
long elliptical shape, being from 6 to 8 inches in width ;
it is composed of an open-worked reticulation of coarse
black or black-brown silk threads, with round or broad
oval interstices ; the fabric is extremely strong, tough,
and elastic, covered externally with moss and birch
leaves firmly adherent.

About a week or ten days before the time of
emergence, the cocoon is pushed by the enclosed pupa
from a prone to a vertical position, the upper end is
ruptured, and the pupa protrudes its head through the
opening, and continues by degrees to advance, until it
is exposed as far as the end of the wing-covers ; fixed
in this position, it remains quiet a longer or shorter
time till the insect is able to escape, though in two or
three instances the pupa had worked itself out entirely
free from the cocoon before the moth could be disclosed ; on examination, the pupa could be seen to be
well furnished with means for facilitating such move-
ments as described below.

The pupa itself measures in the male a length of 12
to 14 or 15 lines, in the female from 17 to 18 lines, or
occasionally a little more ; it is very stout, the diameter
across the bulkiest part, at the end of the wing-covers
in the male, ranges from 4 to 4½ lines, in the female 6
lines. The head has the mouth-parts a little produced
in a squarish form, flanked by the curved antenna-
cases in high relief ; thence the head is bluntly rounded
above in an unbroken swelling curved outline to the
end of the wing-covers, including the thorax and upper
abdominal rings; the movable abdominal ring is very
deeply cut, and below those are well defined, the last

ring ending with a prolonged flattened caudal process
tapering a little to the squarish extremity, where it
has a margin of hooks and bristles; the surface is
remarkably dull, and rough everywhere, except in the
divisions between the movable rings, yet even there
it is quite dull; the roughness on the head, thorax,
upper rings and wing-covers is striated, granulous, or
wrinkled; the movable and lower rings of the abdomen
have on the back transverse rows of stout and sharp
hooks pointing behind; the colour is a sooty or dingy
brown, black in the abdominal divisions. (W. B., 18,
6, 83; E.M.M. XX, 73.)

DREPANA SICULA.

Plate LII, fig. 4.

I feel extremely obliged to Messrs. W. H. Grigg and
W. J. Thomas for giving me the opportunity of
figuring and describing the larvæ of this rare species,
for although it had been described and figured before,
and the description in Stainton's manual is correct as
far as it goes, and one of Hübner's two figures is also
correctly drawn, yet as the other of his figures really
representing D. falcataria has been reproduced under
a wrong name in a recently published book on moths,
the importance of a true representation and description has become all the greater.

Mr. Grigg first sent me an egg which he had obtained June 7th, 1884, from one of two captured
females, kept alive for three days. This egg must
have been fertilised, for during the next five days it
went through the first changes of colour, but finally it
shrivelled up.

Last year, 1876, Mr. Thomas sent me five eggs, laid
June 19th by a pinned moth. These eggs, which
reached me June 23rd, were deposited, three of them
in a little group on a piece of paper, and the other
two loose.

From one of the eggs on the paper the first larva appeared at 11 p.m. June 28th, and a second from one of the loose eggs during the night of June 30th, another of the eggs on the paper never changed colour, and the third, together with the second loose egg, after going through the changes of colour, dried up. Thus I was not very fortunate with the eggs, but most unhappily the young larvæ, after inspiring me with a grand hope, caused me a worse disappointment.

The first that had appeared was supplied within twenty minutes of its emergence with a tender leaflet and a mature leaf of the common lime (*Tilia europæa*); but when I looked at it again, that is, on the following morning, it was dead.

Thinking that perhaps I had failed with the first from not giving it time to eat its egg-shell, when the second larva was hatched I took care to let the empty shell remain with it, and supplied it also with some birch as well as lime leaves; the next day, however, it was looking very miserable, unable to stand, and rolling about helplessly, and so lingered till its death on the third morning, having, as far as I could see, eaten nothing whatever since its hatching.

Why I failed so totally I cannot explain [the explanation appeared three years later, see p. 72, H. T. S.], and can only hope that further experiments with the egg may prove more successful.

Fortunately, however, Mr. Thomas, on September 10th, 1875, had found a nearly full-grown larva on lime, and lent it to me on condition of my sending back the imagos if reared, and this I am happy to say I was able to do. I received the larva September 13th, it became full-fed by the 21st, next day began to spin, and the day after was covered in so as to be hidden, and the moth, a male, and quite perfect, appeared during the evening of the 12th of June, 1876, the first British specimen reared in captivity.

The egg in shape is roundish-oval, the surface very finely pitted; its colour when first laid is pale straw-

yellow, changing in four days to pink at one end, and
a little round the circumferent margin, the centre
remaining straw-colour; on the fifth day the pink
parts turn to the rich red of a ripening strawberry,
and on the ninth the whole surface becomes purplish-
brown.

The newly-hatched larva shows a little of the pecu-
liarity of the adult form, as very slight rudiments of
tubercles can just be detected, and the hinder segment
bears no legs, and is carried at a slight elevation; the
colour is a reddish chocolate-brown, with a darker
brown spot on each lobe of the head. After this point
I can say nothing till the larva is nearly full grown; at
that time I noticed that it span many silken threads to
keep its food steady and to secure its own foothold, and
that its manner of eating was to take large pieces out
from the edges of the lime leaves. At times it rested
with the head and both the anterior and posterior
segments of the body elevated, holding on to the leaf
only by the ventral prolegs, but when walking the
whole of the segments were carried in a tolerably
level line, merely undulating a little in its progress,
though the anal segment had always a slight upward
turn.

The full-grown larva measures one inch in length,
and is in proportion moderately slender, with fourteen
legs, the anal segment not having legs, but is much
prolonged to a tapering point curving a little upwards;
the head much larger than the second segment, and
broadest near the mouth, the crown and lobes erect
and deeply cleft and flattened in front, as is the whole
face; on the back of the fourth segment is an elevated
process, divided into two blunt-tipped tubercles
(Hübner's figure above referred to has *four* pairs of
tubercles); the segments generally are moderately
defined above, more deeply below, and very delicately
wrinkled, with three or four subdivisions across the
back of each; the head is pinkish flesh colour, and a
clear margin of this is left on the ridges of the divided

crown, whence it is relieved below by brown spots on the face, and by a dark brown outline of the lobes there, and on the middle of each lobe are three lines of dark spots, *i. e.* one spot at the side of the cheek, two in the middle, and a few more very minute between and above them, also a dark spot or two about the mouth, the papillæ whitish, the ocelli black, the thoracic segments much suffused with brownish-ochreous, on which both the dorsal and subdorsal regions are strongly blotched with dark crimson-brown, the tubercles brown with yellow tips; hence the colour of the back and portions of the sides is a brilliant deep yellow, bearing extremely minute elon-gated freckles of a dark brown, a series of these freckles faintly indicating dorsal and somewhat of sub-dorsal lines, and an assemblage of them close together constitute a dark spot on each side of the twelfth segment near to the only distinctly noticeable spiracle, which is there seen as a faint brown oval outline; a fine hair proceeds from each of the usual localities; on the belly, the legs, and some portions of the sides the ground colour is pink, deeply tinged and freckled above with dark crimson-brown; this fluctuates along the middle segments of the body in two distinct waves on either side from the spiracular region of the fifth segment, and falls again rather lower each time than its previous level, till at the eleventh segment and onward to the anal point it covers scarcely more than the ventral surface; the summits of these dark wavès reach high on the back of the sixth and ninth seg-ments in such strong contrast to the yellow as to create something of an optical illusion as to the shape of the body.

When the larva prepared for changing it began to spin upon the upper surface of a leaf not very far from the foot-stalk, and soon contrived to draw upwards a portion of the two sides so as to form a cavity, to which the midrib of the leaf would be a support below, though its actual position was not quite in the

middle of it ; the walls (so to speak) in a short time
began to approach each other as the foundations were
progressing, which consisted of three or four thick
little pads of silk attached on either side opposite each
other, drawn from time to time closer and closer
together and connected by very short and stout
threads; these were presently rendered still shorter
by a few threads drawn amongst them from one to
the other at various angles which seemed to contract
the opening, and to bring the stud-like fastenings
almost together. At this point the larva rested for
awhile, and next morning I found any further watching was effectually prevented; the cocoon had been
made, and the covering of the narrow orifice last spun
was of a pale, rusty-red colour, which in a day or two
became a little darker, while the earlier fastenings I
had seen spun turned purplish-brown, looking like
dark veins on a decaying part of a leaf.

On examining the cocoon after the exit of the moth,
I found it very smooth within and made of very strong
silk, not easily torn open, of a rusty-brown colour
like that of the leaf containing it ; the pupa-skin presented but little on which to remark in its general
form, its length a little over half an inch and its
diameter three-sixteenths; in figure it was rounded at
the head, and tapering but little till near the anal tip,
which was rather prolonged in a blunt point ending
with three central curled-topped bristles, with a
shorter one on each side of them; where the thorax
and another part had been rubbed a little, the rusty-
red colour of the pupa-skin without gloss and rather
rough, could be seen. The rest of the surface was
covered with a soft adhesive opaque white powder.
(W. B., 5, 5, 77; E.M.M. XIV, 1.)

I am now able to offer a few more observations to
fill up the hiatus in the early part of the history of
this larva, and am much indebted to Mr. W. H. Grigg
for his perseverance and kind help, which have enabled
me to give the following details :

The eggs are laid by the parent moth on the very edges of the leaves, so that when hatched her progeny shall find themselves exactly where their food is most suitable, for however much they may wander at first, it is there, in preference to any other part, the young larvæ invariably begin to feed on the cuticle of the upper surface; there also they spin a small quantity of silk, on which to rest and be secure while moulting.

After a moult, while the larva is but little more than one-eighth of an inch long, the future form is indicated, though the segments are strongly wrinkled and folded across at intervals, and the previous plain, chocolate-brown colour is exchanged for russet-brown, relieved by minute dots and transverse bars of yellow.

After the next moult, the larva begins to cut quite through the substance of the leaves, eating out semi-circular portions from the edge; it also begins to show on the brown-ground colour little patches of very subdued ochreous yellow in angular forms on the back. Five days later it spins a quantity of silk, tying as it were the leaves loosely together, but firmly, for its safety while laid up for another moult, which is accomplished after two or three more days, and then it has the characteristic party-coloured coat of dark velvety-brown and pale cream colour, the tubercular process appearing on the fourth segment as two short black eminences. It soon spins more silk threads, keeping the leaves partly together, and feeds well until once more laid up for moulting, and this takes place about a week after the previous change of skin.

Now the rosy-pink colour appears on the belly and ventral prolegs, and the yellow parts of the back have a thin brown dorsal and fine lateral lines, the yellow being much brighter than before; three distinct shapes of yellow are seen on the back, well defined, and contrasted by the rich dark brown surrounding them, viz. a brilliant pale yellow triangular mark, its base at the beginning of the fifth segment, its apex at the beginning of the sixth; an elongated diamond

mark of deeper yellow extends from near the beginning of the seventh segment to near the end of the
ninth; another begins on the front of the tenth and
includes the pointed tail, relieved on the twelfth
segment with a brown chevron. As the larva grows
these yellow marks expand and become united into
one long fluctuating shape along the back, as I have
formerly described, though I have since then had one
variety retain the triangular mark isolated distinctly to
the end of its larval existence; and another with the
yellow colour rather inclining to drab.

In reference to my former account of the species
(p. 67), wherein mention was made of two young
larvæ dying rather than eat the lime supplied to them,
and that yet only the year before a nearly mature
larva had thriven on that food well enough, it is
now needful to state that what seemed to me then so
inexplicable, received afterwards an easy solution
when Mr. Grigg sent me some lime gathered in the
haunts of *D. sicula*, leaves whose smaller size and
qualities of texture and colour were different from
those the little larvæ rejected. It was a great satis-
faction, then, on visiting the trees where, without
thought of any particular species of lime, I had first
gathered food for the adult larva, to find that they
were *Tilia parvifolia*, and that *T. europæa* also grew
at no great distance, to which, by a mischance, the
next year at night my footsteps had been directed, an
incident proving the importance of having the proper
name when allusion is made to trees or plants as food
for larvæ. (W. B., 10, 10, 80; E.M.M. XVII, 122.)

DREPANA HAMULA.

Plate LIII, fig. 2.

On the 27th of August, 1879, I received from Mr. W. R. Jeffrey a dozen eggs laid by a captive female; they were laid between the 18th and 20th of August on the edges of leaves of oak, here and there one on the very edge of a leaf.

At first they were greenish-white, but in the course of two or three days they changed to a light fawn colour, and afterwards to a bright deep red at each end of the shell, with an irregular blotch of fawn colour glistening in the mid-surface.

In shape the egg is oval, with a depression on its upper surface and ribbed longitudinally.

On the 30th two larvæ were hatched, and two more on the 31st, one of which died soon after without feeding (of the remaining eggs some did not change colour, others turned red, but afterwards shrivelled up).

The newly-hatched larva carries its hinder segments a little elevated; it is brown, with the head paler. On the 7th September the first moult was over. (W. B., Note Book III, 289.)

HETEROGENEA ASELLUS.

Plate LIII, fig. 7.

On the 13th October, 1872, I had the good fortune to receive an example of the larva of this species, found on a beech tree near Marlow, and kindly presented to me by the Rev. B. Smith.

For two days the larva continued to feed at intervals on the edges of beech leaves, and on the 16th it spun its cocoon on the under surface near the edge of one of the leaves, and the perfect insect, a female, came forth on the 7th July, 1873.

This larva, when moving and fully stretched out,

measured about half an inch long, and a quarter of an inch broad across the middle of the body, whence it tapered towards each end; but in repose, or when disturbed, its length did not exceed three-eighths of an inch, as the head and the second segment were then entirely retracted, so that the front part of the body appeared but little tapered, and broadly truncate, though somewhat rounded. When protruded, the head was seen to be very small, and rather flattened as in the *Lycænidæ;* the anal extremity was rounded; viewed sideways the back appeared somewhat arched, and the ventral surface was in close contact with the leaf. The segments were not marked in the usual way by transverse folds, but only by narrow dimpled depressions; there were also little circular dimples on the back, one in the centre of the front of each segment, and two at the back in the subdorsal region; this region, being a little raised on each side, formed a slight dorsal hollow.

The six anterior legs, though minute, were yet distinctly to be seen when the larva was in motion, but no ventral or anal legs were perceptible, and instead of them it had along the sides on the margin of the belly, which was deeply depressed along the middle, a soft projecting ridge of extremely flexible skin. This served very well the purpose of legs by its undulatory movement from behind forwards; one wave at a time, formed under each segment, slowly advancing and subsiding in regular succession as far forwards as the fifth segment.

The head was smooth and shining, the back and sides rather so, though the skin there was covered with a pubescence, but this was so fine as to be seen only with a powerful lens; it was noticeable that the dimpled spots were for the most part paler than the rest, and that a few short and very minute bristles were scattered at each extremity, and at intervals along the back.

The ground colour was a pale yellowish-green, watery looking along the sides, where it soon faded into something of a pale flesh tint beneath; on the back, beginning at the front of the third segment, was a broad, olive-brown, extensive mark, reminding one somewhat of the dark saddle on *Cerura vinula;* this mark lessened in breadth a little on the fifth segment, and then grew broader on the sixth, attaining its greatest breadth on the seventh and eighth, where it reached low on the sides; it began to decrease again on the ninth and thence gradually narrowed to the anal tip. The olive-brown was darkest on the third and fourth segments, and there was throughout an outline of darker brown; this was further relieved below by a pale, sulphur-yellow border, which enlarged to a spot on the side of the fifth segment, with smaller spots on the fourth and sixth; the second segment was pale yellowish-green, and also the head, with a slight tinge of brown, the mouth edged above and on each side with dark brown, papillæ yellowish-green. A dark brown dorsal vessel could be seen through the olive on the back as far as the end of the tenth segment; the moving skin of the under surface was almost colourless, with a clear, pellucid, jelly-like appearance.

The cocoon was a quarter of an inch long, two lines broad, of a very short elliptical form, bearing a great resemblance to a gall excrescence; a few fine threads formed a kind of network round its base attached to the leaf; it was dark dull brown, with blotches of a pale grey film spreading irregularly over the upper surface, as though it bore a delicate lichenous growth.

The pupa-skin when extracted was but a trifle less than a quarter of an inch long, and thick in proportion, the abdomen bent under, which gave it rather a rounded form; the abdominal segmental divisions were distinct, as were the parts of the head and thorax; the wing cases were well developed and projecting; it was all very smooth and polished, and of a transparent whity-brown colour. (W. B., 12, 7, 73; E.M.M. X, 70.)

On the 13th July (1877) the Rev. B. Smith, of

Marlow, kindly sent me a few eggs, procured by Mr. Felix Parker, together with the parent female, which had a small bunch of eggs protruding from the ovipositor; the others were laid on the side of a chip box, agglomerated together.

Viewed with a lens they appeared somewhat of a drop shape, but ill defined, from their being connected together in little lumps, the colour very pale, shining, transparent, and gelatinous-looking, otherwise much the colour of the chip. By the end of the month they began to grow yellowish, and then to be tinged with the colour of brown sherry in parts of the little masses; then they began to hatch. At this critical moment, being otherwise engaged for some hours, I was unable to know that they were hatched and requiring food, though I was keeping them in a glass-topped box in order to observe the changes of colour. When I returned to them all were dead except one, a mere speck, which was slowly moving on the chip; one dead one lay at the bottom of the box, and others I found dead, wedged into the junction of the box and its lid. As well as my strongest lens would show them to me, these very small specks of creatures were of an ovate roundish figure, dark brown above and pale greenish beneath,—in short, miniature representations, apparently, in all respects of the mature larva.

The solitary living larva I placed on a leaf of beech, and put two other leaves over it, but on looking for it two days later was unable to see it, and concluded it had somehow escaped, probably through the muslin cover of the little perforated box in which it was confined. (W. B., Note-Book III, 200.)

The following list of parasites, bred from the larvæ or pupæ of the species included in the present volume, has been kindly prepared by Mr. G. C. Bignell, F.E.S. —H. T. S.

HOST.	PARASITE.	By whom bred.
Diloba cæruleocephala	*Mesochorus formosus*, Bridgman	Miss N. P. Decie.
	Apanteles difficilis, Nees	Decie.
	Apanteles fulvipes, Haliday	G. C. Bignell.
Clostera reclusa.........	*Pimpla graminellæ*, Schrank	W. H. B. Fletcher.
Pygœra bucephala......	*Campoplex mixtus*, Fabricius	Bignell, T. A. Marshall, F. Norgate.
	Campoplex falcator, Thunb.......	Norgate.
	Pimpla instigator, Fabricius ...	E. A. Fitch.
Dasychira fascelina ...	*Campoplex ebeninus*, Gravenhorst	Bignell, W. H. Harwood.
,, *pudibunda*.	*Trogus alboguttatus*, Gravenhorst	Bignell.
	Limneria unicincta, Gravenhorst	Bignell.
	Apanteles triangulator, Wesmael	Bignell.
	Exorista gnava, Meigen	Bignell.
Demas coryli	*Schizoloma amicta*, Fabricius ...	Marshall.
Orgyia antiqua	*Limneria obscurella*, Holmgrén .	W. Bennett.
	Mesochorus stigmaticus, Thompson	Norgate.
	Apanteles solitarius, Ratzeburg	Bignell.
Stilpnotia salicis	*Pimpla instigator*, Fabricius ...	Fitch.
	Pelecystoma lutea, Nees	Fitch.
	Apanteles salebrosus, Marshall	Bignell, G. F. Mathew.
Porthesia chrysorrhœa	*Pimpla instigator*, Fabricius ...	J. J. Weir.
,, *auriflua*......	*Pimpla instigator*, Fabricius ...	Fitch.
	Apanteles fulvipes, Haliday	Bignell.
	Microgaster posticus, Nees	Bignell.
	Microgaster connexus, Nees	Bignell.
Lithosia complanula...	*Microgaster globatus*, Linné......	B. A. Bower.
Œnistis quadra.........	*Pimpla instigator*, Fabricius ...	T. Eedle.
Gnophria rubricollis...	*Eurylabus torvus*, Wesmael ...	Marshall.
Nudaria mundana ..	*Casinaria mesozosta*, Holmgrén	C. G. Barrett.
Nola albulalis	*Limneria Fitchii*, Bridgman	Bignell, W. H. Tugwell.
Arctia caja..............	*Pimpla instigator*, Fabricius ...	Fitch.
	Ascogaster rufidens, Wesmael ...	Fitch.
	Apanteles caiæ, Bouché.........	Bignell, W. J. Cross.

* Hyperparasite on *Apanteles difficilis*.

Host.	Parasite.	By whom bred.
Arctia caja (continued)	Apanteles difficilis, Nees	Fitch.
	Macrocentrus pallipes, Nees	W. P. Weston.
	Baumhaneria vertiginosa, Meigen	Fitch.
	Thelaira leucozona, Panzer	E. A. Butler.
„　villica	Apanteles caiæ, Bouché	Eedle.
Nemeophila plantaginis	Apanteles callidus, Haliday	Bignell.
Phragmatobia fuliginosa	Ichneumon haglundi, Holmgrén	T. R. Billups.
	Limneria interrupta, Holmgrén	G. H. Raynor.
Spilosoma menthastri	Ichneumon sarcitorius, Linné {	J. Hellins, Bignell.
	Apanteles ruficrus, Haliday	Bignell.
	Apanteles nothus, Reinhard	Bignell.
Callimorpha jacobææ	Cryptus migrator, Fabricius	J. Sang.
	Cryptus incubitor, Strœm	Sang.
	Anomalon cylindricum, Bridgman	Fitch.
	*Mesochorus facialis, Bridgman	Bignell.
	Apanteles difficilis, Nees	Bignell.
	Apanteles popularis, Haliday {	Bignell, Butler, Marshall, R. M. Sotheby.
Lasiocampa rubi	†Apanteles difficilis, Nees	Bignell.
„　trifolii	Gravenhorstia picta, Boie	R. Mitford.
„　quercus	Cryptus migrator, Fabricius... {	Bignell, Fitch.
	Ophion ventricosum, Gravenhorst	Marshall.
	Limneria rufa, Bridgman	Bignell.
	Bassus nigritarsus, Gravenhorst	Bignell.
	Metopius micratorius, Fabricius	Mrs. Norgate.
	Thryptocera bicolor, Meigen	Bignell.
Eriogaster lanestris	Eurylabus dirus, Gravenhorst {	Bignell, Marshall, Sotheby, T. Wilson.
	Phæogenes calopus, Wesmael	Bignell.
Pœcilocampa populi	Apanteles difficilis, Nees {	J. B. Robson, Bignell.
Clisiocampa neustria	Exorista vulgaris, Fallén	Norgate.
Odonestis potatoria	Pimpla graminellæ, Schrank	Fitch.
	‡Rhogas geniculator, Nees {	Bignell, Fletcher, Mathew.

* Hyperparasite on *Apanteles popularis*.
† Emerging from the young larva before the third moult.
‡ The perfect fly emerges from the infested larva while it appears to be preparing for the fourth moult.

Host.	Parasite.	By whom bred.
Odonestis potatoria (continued)	*Exorista vulgaris*, Fallén	S. Mosley, G. T. Porritt.
Saturnia carpini	*Amblyteles armatorius*, Forster .	Marshall.
	Cryptus fumipennis, Gravenhorst	Barrett, J.B.Bridgman A. Elliot. G. Elisha, Fitch, R. Meldola.
	Pezomachus insolens, Gravenhorst	Barrett, Elliot.
	Apanteles immunis, Haliday ...	Decie.
	Exorista grandis, Zetterstedt .	Bignell, Fitch, Mosley.
Platypteryx lacertinaria	*Apanteles sericeus*, Nees............	Bower.
Drepana falcataria ...	*Macrocentrus linearis*, Fabricius	Bignell.
Heterogenea asellus ...	*Sagaritis declinator*, Gravenhorst	Bignell.
	Limneria unicincta, Gravenhorst	Bignell.
Limacodes testudo ...	*Pelecystoma lutea*, Nees	Fletcher.

INDEX.

PLATE XXXVI.

Diloba cæruleocephala.

1, young larva after first moult; 1 *a*, larva after second moult; 1 *b*, 1 *c*, larvæ after third moult; 1 *d*, larva after last moult.

See pp. 1—4.

Petasia cassinea.

2, 2 *a*, larva after last moult.

Petasia nubeculosa.

3, 3 *a*, larva after last moult. Figure 3 is rather too long and the back too straight. (W. B.'s MS.)
See pp. 4—9.

Peridea trepida.

4, larva before last moult; 4 *a*, larva after last moult.
See pp. 9, 10.

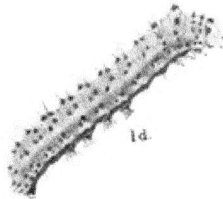

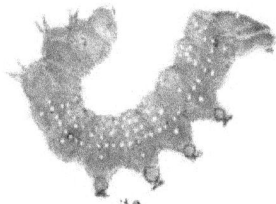

PLATE XXXVII.

CLOSTERA RECLUSA.

1, larva after last moult.

CLOSTERA ANACHORETA.

2, 2 *a*, larvæ after last moult.

CLOSTERA CURTULA.

3, larva after last moult.

PYGÆRA BUCEPHALA.

4, 4 *a*, larvæ after last moult.

PSILURA MONACHA.

5, larva after last moult; 5 *a*, pupa.

HYPOGYMNA DISPAR.

6, 6 *a*, larvæ after last moult.

1

2

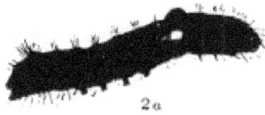

2a

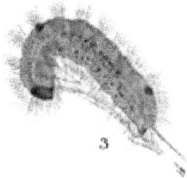

3

4

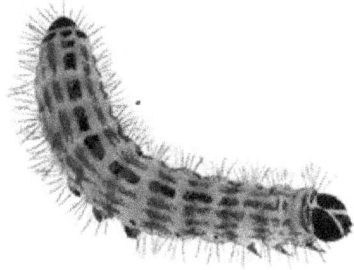

4a

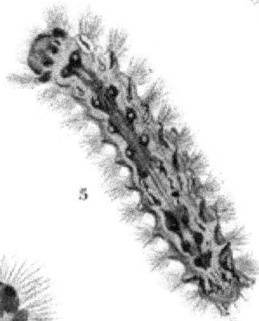

5

5a

6

6a

PLATE XXXVIII.

Dasychira fascelina.

1, 1 *a*, larvæ after last moult.

Dasychira pudibunda.

2, 2 *a*, 2 *b*, 2 *c*, 2 *d*, 2 *e*, larvæ after last moult (2 is exceptionally small and 2 *d* unusually large, H. T. S.); 2 *f*, cocoon; 2 *g*, pupa.

Demas coryli.

3, young larva; 3 *a*, 3 *b*, 3 *c*, 3 *d*, 3 *e*, 3 *f*, larvæ after last moult.

Plate XXXVIII

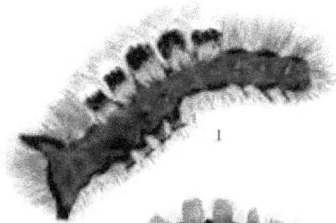

1

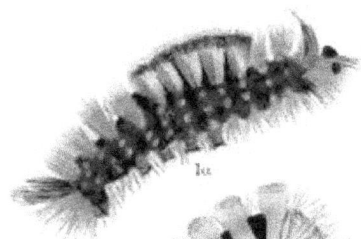

1a

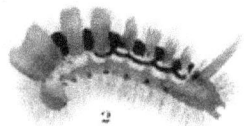

2

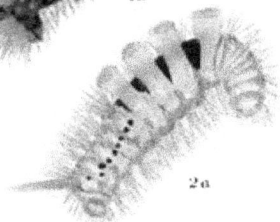

2a

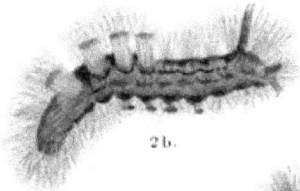

2b.

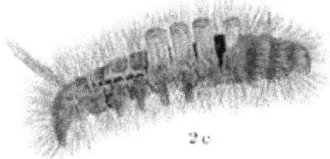

2c

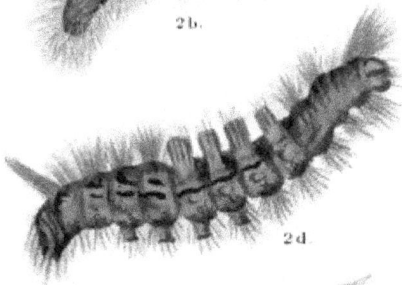

2d

2e.

2g.

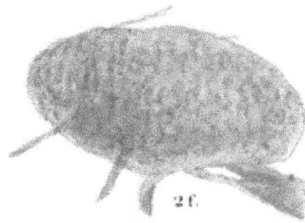

2f.

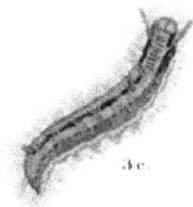

3c.

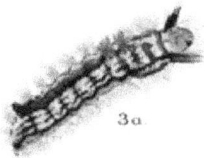

3a.

3

3d.

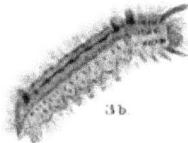

3b.

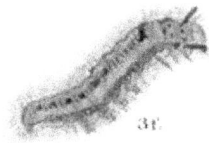

3f.

3e.

F C Moore lith.

West, Newman & Co. imp.

PLATE XXXIX.

ORGYIA ANTIQUA.

1, 1 *a*, 1 *b*, larvæ after the last moult; 1 *c*, cocoon with the eggs deposited by the wingless female.

See pp. 11, 12.

ORGYIA GONOSTIGMA.

2, 2 *a*, young larvæ as found in autumn; 2 *b*, larva after last moult, end of May or beginning of June.

LÆLIA CÆNOSA.

3, larva after last moult.

STILPNOTIA SALICIS.

4, 4 *a*, larvæ after last moult.

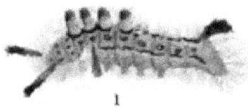

1

1a

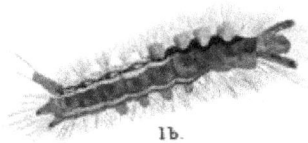

1b.

1c

2

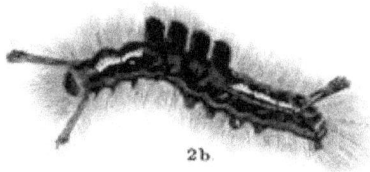

2b

2a.

3

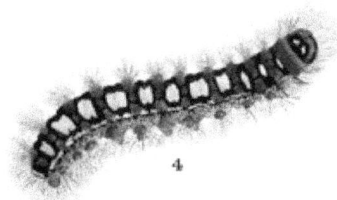

4

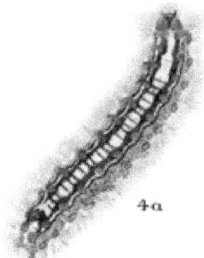

4a

PLATE XL.

PORTHESIA CHRYSORRHÆA.

1, 1 *a*, larvæ after last moult.

PORTHESIA AURIFLUA.

2, larva after last moult; 2 *a*, cocoon.

MILTOCHRISTA MINIATA.

3, 3 *a*, 3 *b*, larvæ after last moult.
See pp. 13, 14.

LITHOSIA CANIOLA.

4, larva after last moult, from green lichens on walls, &c.; 4 *a*, larva after last moult, from flowers of *Lotus corniculatus*, Howth, Ireland.
See pp. 14, 15.

LITHOSIA AUREOLA.

5, 5 *a*, larvæ after last moult.
See pp. 15, 16.

LITHOSIA HELVOLA.

6, 6 *a*, 6 *b*, larvæ after last moult.
See pp. 16, 17.

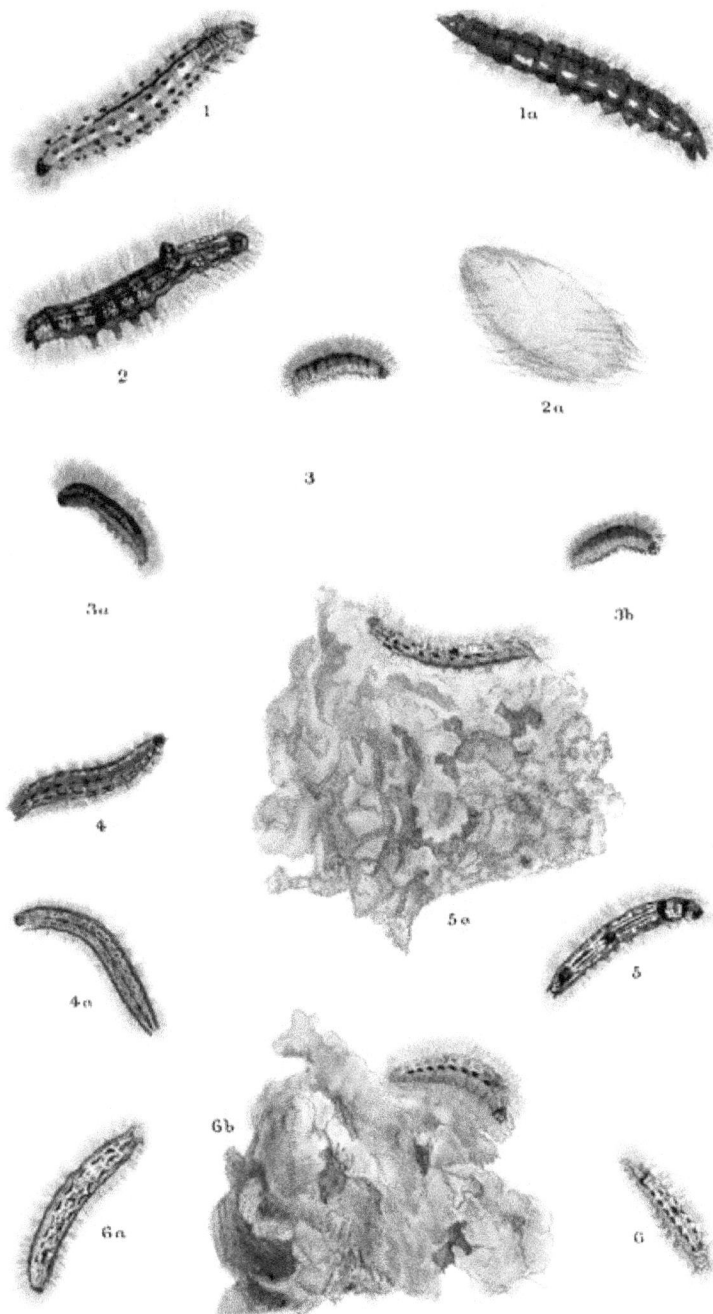

Newman & C° imp.

PLATE XLI.

Lithosia stramineola.

1, 1 *a*, 1 *b*, larvæ after last moult.
See pp. 17—19.

Lithosia complana.

2, 2 *a*, 2 *b*, 2 *c*, 2 *d*, larvæ after last moult.
See pp. 19—21.

Lithosia complanula.

3, 3 *a*, larvæ before last moult; larva after last moult.
See pp. 21, 22.

Lithosia molybdeola.

4, 4 *a*, 4 *b*, larvæ after last moult.
See pp. 22, 23.

Lithosia griseola.

5, 5 *a*, 5 *b*, 5 *c*, 5 *d*, larvæ after last moult.
See pp. 23, 24.

Lithosia pygmæola.

6, larva after last moult.
See pp. 25, 26.

Lithosia muscerda.

7, 7 *a*, 7 *b*, larvæ after last moult.
See pp. 26—28.

Plate XLII.

ŒNISTIS QUADRA.

1, arva before last moult; 1 *a*, 1 *b*, 1 *c*, larva after last moult.

See pp. 29—33.

GNOPHRIA RUBRICOLLIS.

2, larva before last moult; 2 *a*, larva after last moult.

See p. 34.

CYBOSIA MESOMELLA.

3, 3 *a*, young larvæ before hibernation; 3 *b*, 3 *c*, larvæ after last moult, in May.

See pp. 34, 35.

PHILEA IRRORELLA.

4, two larvæ after last moult feeding on small blackish lichens on stones of sea-beach above high-water mark (at Portslade, near Brighton).

See pp. 36, 37.

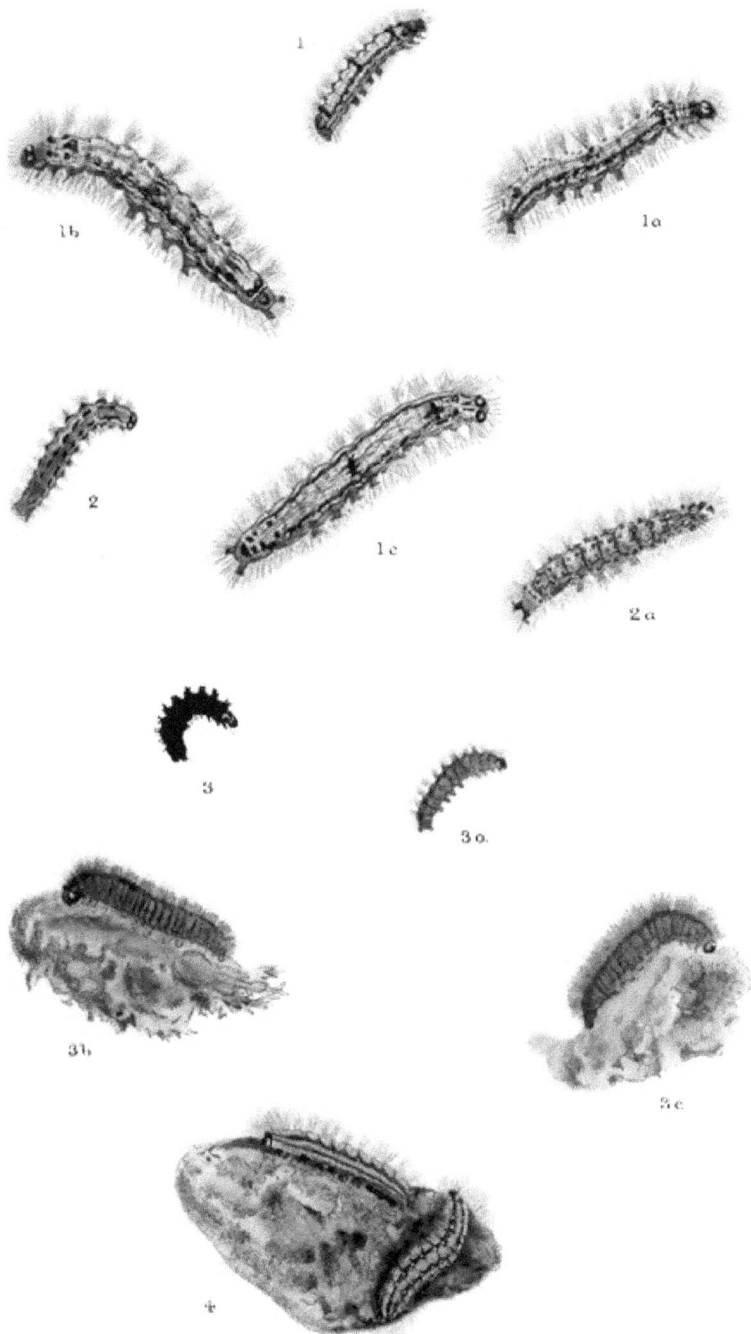

1

1 b

1 a

2

1 c

2 a

3

3 a.

3 b

3 c

4

F C Moore lith   W BUCKLER del   West, Newman & Co. imp

PLATE XLIII.

Nudaria mundana.

1, 1 *a*, larvæ after last moult.

See p. 37.

Nudaria senex.

2, 2 *a*, 2 *b*, 2 *c*, 2 *d*, larvæ after last moult; 2 *e*, cocoon or pupa-case.

See pp. 37—39.

Nola cucullatella.

3, larva after last moult.

Nola strigula.

4, 4 *a*, 4 *b*, larvæ after last moult.

See pp. 39—41.

Nola albulalis.

5, 5 *a*, 5 *b*, larvæ after last moult; 5 *c*, cocoon attached to a stem.

See pp. 41—43.

Nola centonalis.

6, 6 *a*, larvæ before last moult; 6 *b*, larva after last moult; 6 *c*, cocoon attached to a stem.

See pp. 44—50.

Nola cristulalis.

7, larva after last moult.

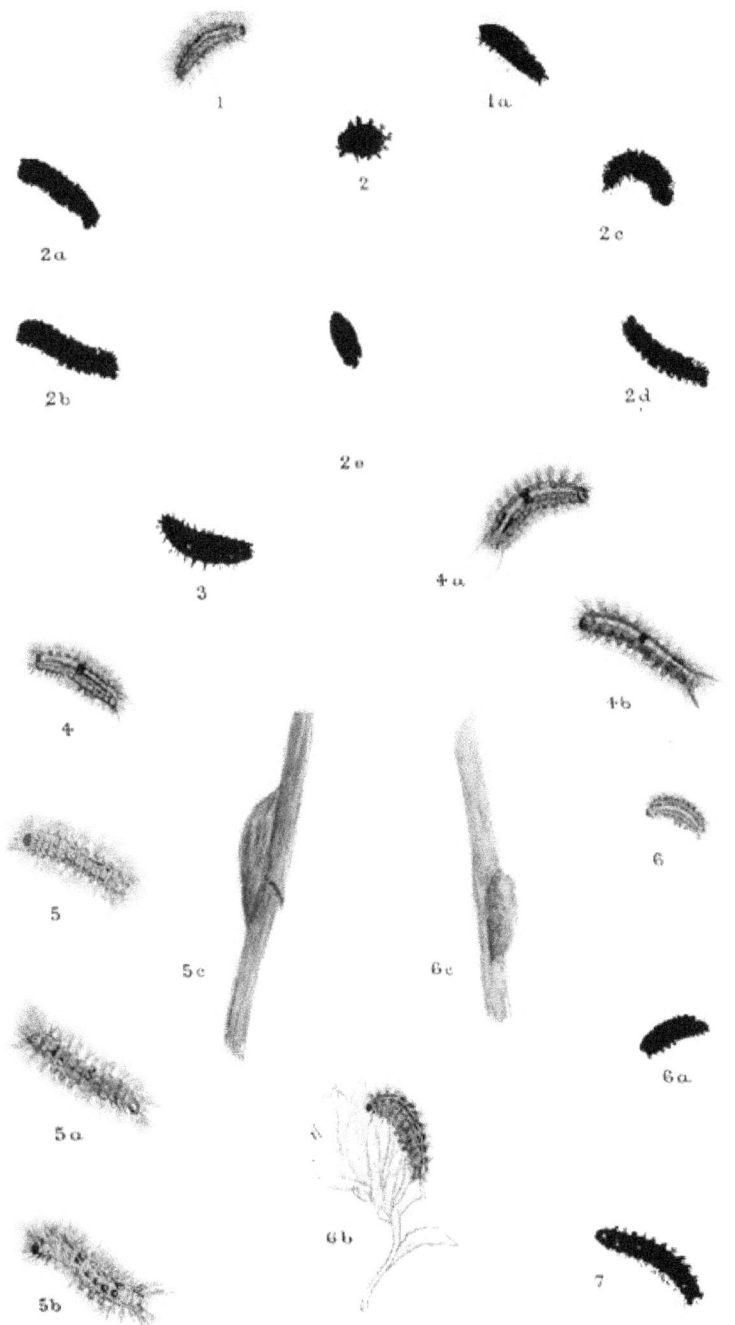

1   1 a

2

2 a   2 c

2 b   2 d

2 e

3   4 a

4   4 b

5   6

5 c   6 c

5 a   6 a

5 b   6 b   7

F.C.Moore lith.   W. BUCKLER del.   West Newman & Co imp

PLATE XLIV.

HYPERCOMPA DOMINULA.

1 larva before last moult; 1 *a*, 1 *b*, larvæ after last moult (1 *a* was copied by Mr. Buckler from a drawing by Mr. Standish).

EUTHEMONIA RUSSULA.

2, 2 *a*, larvæ after last moult; 2 *b*, pupa.

ARCTIA CAJA.

3, young larva. N.B.—No figure of the adult larva by Mr. Buckler which would bear reproduction exists; being one of our very commonest species, it could be figured at any time; moreover, every entomologist is perfectly familiar with the larva, and needs no figure of it.—H. T. S.

ARCTIA VILLICA.

4, larva before last moult; 4 *a*, larva after last moult.

NEMEOPHILA PLANTAGINIS.

5, 5 *a*, 5 *b*, larvæ after last moult.

1

1a

1b

2

2a

3

2b

4

5

4a

5a

5b

PLATE XLV.

Phragmatobia fuliginosa.

1, young larva, this moulted the next day and became brown and very hairy; 1 *a*, larva after last moult.

Spilosoma menthastri.

2, 2 *a*, larvæ before last moult; 2 *b*, 2 *c*, larvæ after last moult.

Spilosoma papyratia.

3, larva after last moult.

Spilosoma lubricepeda.

4, young larva; 4 *a*, 4 *b*, larvæ after last moult.
See pp. 50—52.

Diaphora mendica.

5, larva after last moult.

1

1a

2

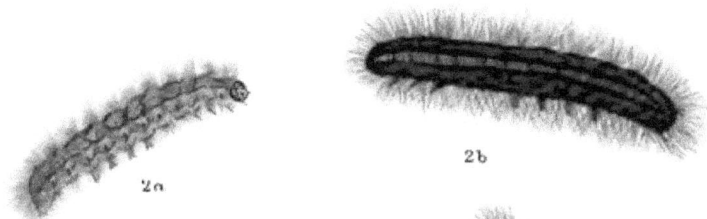

2a

2b

3

2c

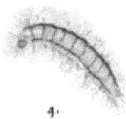

4

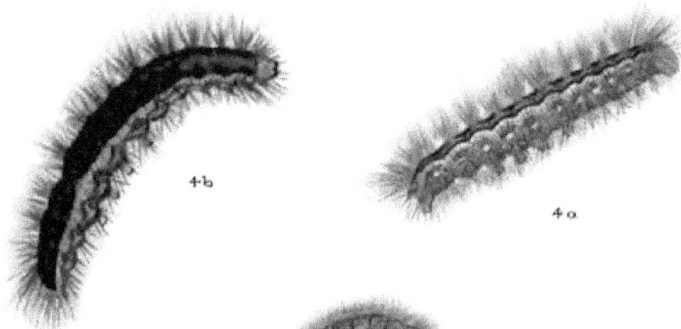

4b

4a

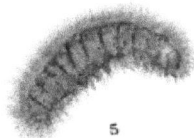

5

F C Moore lith.   W. BUCKLER del   West Newman & Co imp

PLATE XLVI.

Callimorpha jacobææ.

1, 1*a*, larvæ after last moult.

Eulepia cribrum.

2, larvæ after last moult.

Deiopeia pulchella.

3 *a*, 3 *b*, larvæ before last moult; 3, 3 *c*, 3 *d*, larvæ after last moult; 3 *e*, two segments of the adult larva magnified; 3 *f*, pupa.

See pp. 52—56.

Lasiocampa rubi.

4, larva before last moult; 4 *a*, larva after last moult; 4 *b*, pupa; 4 *c*, cocoon.

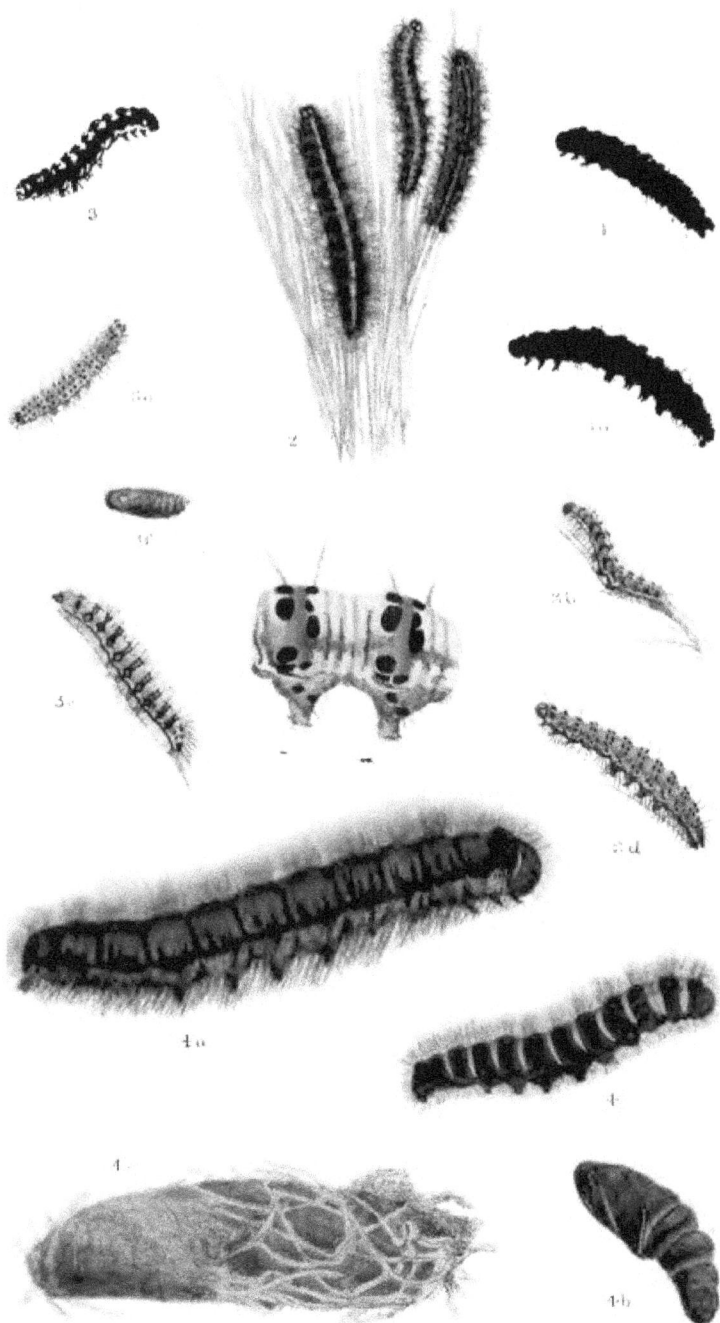

Plate XLVI

F.C. Moore lith.   W. BUCKLER del.   West, Newman & Co. imp.

PLATE XLVII.

LASIOCAMPA TRIFOLII.

1, larva after last moult; 1 *a*, white variety of the adult larva; 1 *b*, cocoon.

LASIOCAMPA QUERCUS.

2, larva before last moult; 2 *a*, black variety of the larva found on sloe; 2 *b*, larva after last moult.
See pp. 56—57.

LASIOCAMPA QUERCUS, var. CALLUNÆ.

3, 3 *a*, 3 *b*, young larva; 3 *c*, larva before last moult; 3 *d*, larva after last moult.
See p. 58.

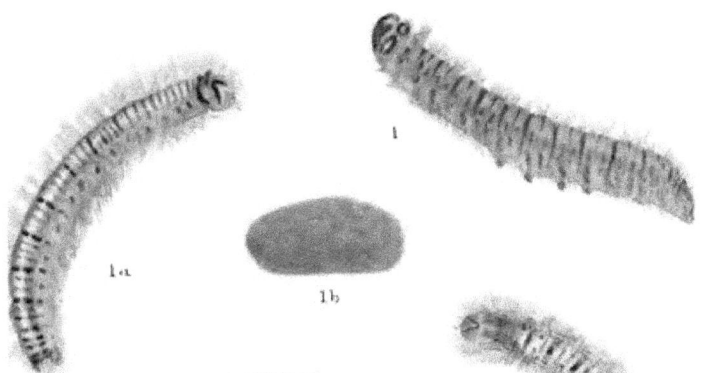

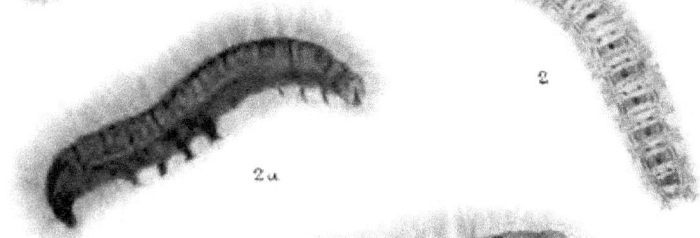

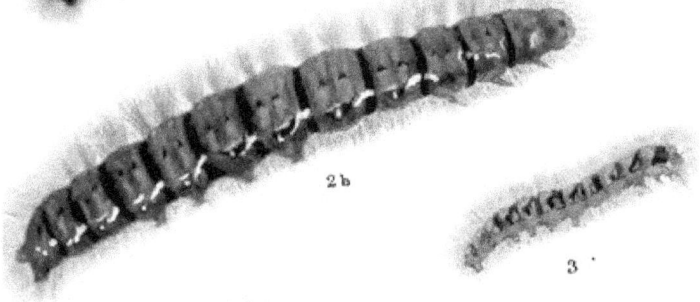

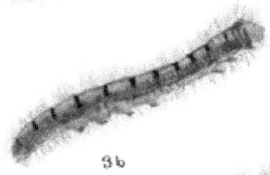

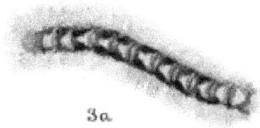

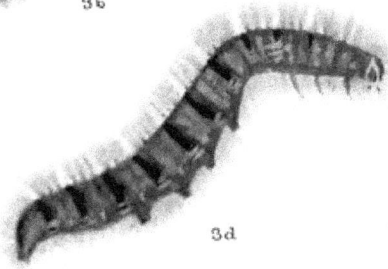

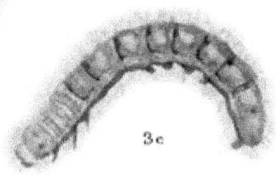

1

1a

1b

2

2a

2b

3

3b

3a

3d

3c

F C Moore lith. W BUCKLER del. West Newman & Co lith

PLATE XLVIII.

Eriogaster lanestris.

1, 1 *a*, larvæ before last moult; 1 *b*, 1 *c*, 1 *d*, larvæ after last moult; 1 *e*, cocoon.

Fig. 1 *d* is just a little too thick along the side below the yellow markings. (W. B.'s MS.)

Pœcilocampa populi.

2 *b*, larva before last moult; 2, 2 *a*, 2 *c*, larvæ after last moult. (See also Plate XLIX, figs. 1, 1 *a*.)

See pp. 58—60.

Plate XLVIII

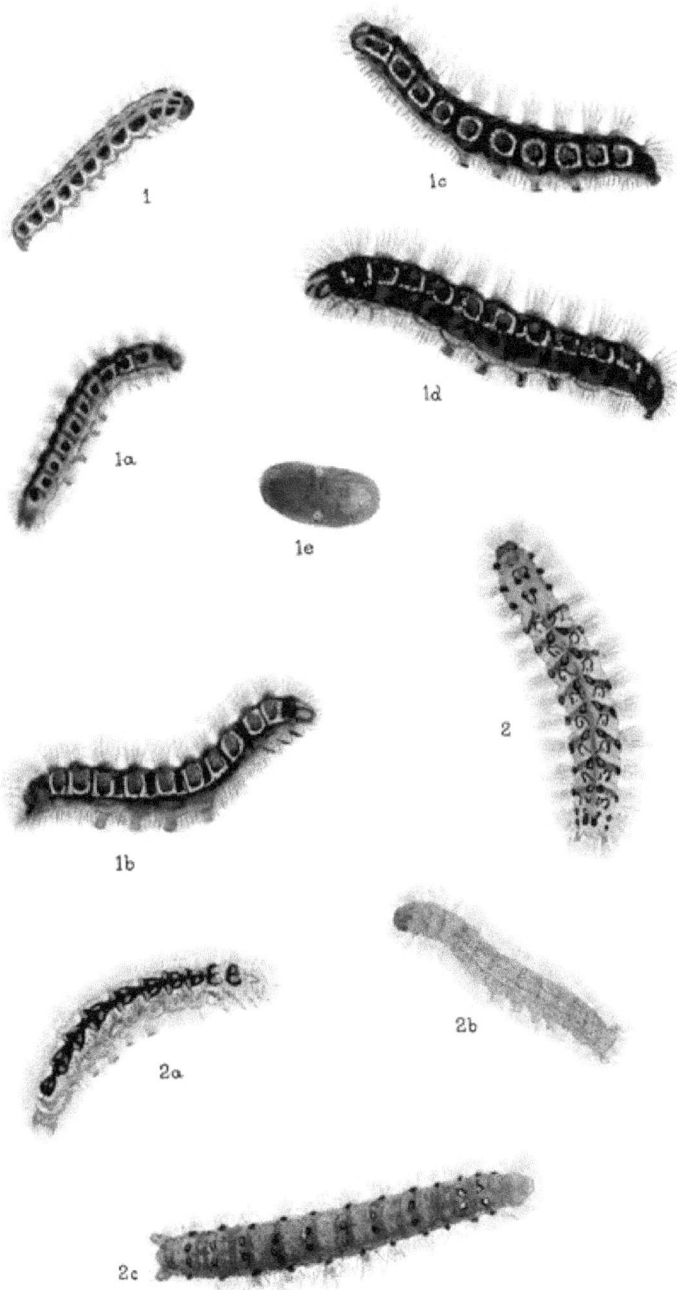

PLATE XLIX.

Pœcilocampa populi.

1, larva after last moult ; 1 *a*, cocoon.   (See also Plate XLVIII, figs. 2, 2 *a*, 2 *b*, 2 *c*.)

See pp. 58—60.

Trichiura cratægi.

2, 2 *a*, larvæ before last moult ; 2 *b*, 2 *c*, 2 *d*, 2 *e*, 2 *f*, larvæ after last moult ; 2 *g*, cocoon.

Plate XLIX.

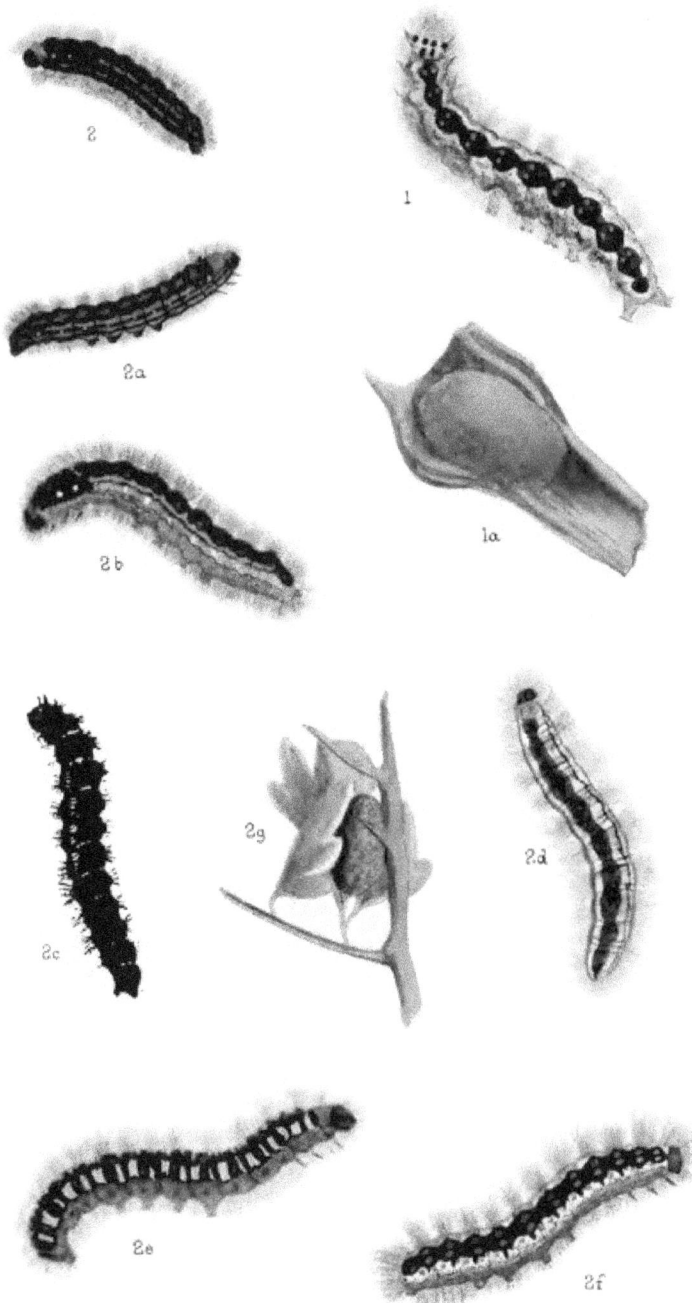

PLATE L.

CLISIOCAMPA CASTRENSIS.

1, 1 *a*, 1 *b*, larvæ after last moult; 1 *c*, cocoon.

CLISIOCAMPA NEUSTRIA.

2, 2 *a*, larvæ after last moult.

ODONESTIS POTATORIA.

3, young larva before hibernation; 3 *a*, 3 *b*, larvæ after last moult; 3 *b* shows the characteristic attitude assumed by the larva when touched.

See p. 60.

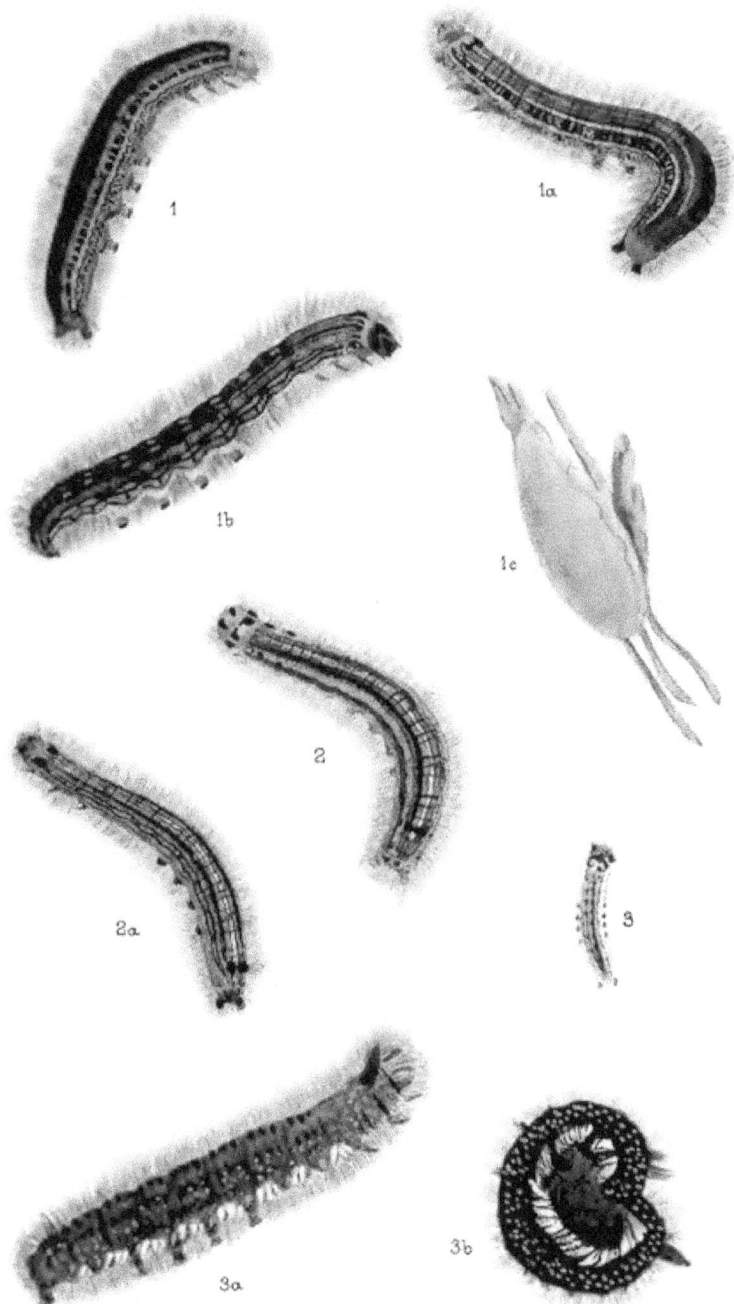

Plate L

1

1a

1b

1c

2

2a

3

3a

3b

F C Moore lith

W BUCKLER del

West Newman & C°.imp

PLATE LI.

Gastropacha quercifolia.

1, 1 *a*, larvæ after last moult ; 1 *b*, cocoon.

Gastropacha ilicifolia.

2, larva after last moult ; 2 *a*, pupa.

Endromis versicolor.

3, 3 *a*, larvæ after last moult;   3 *b*, pupa.
See pp. 60—66.

Plate LI

1

1b

1a

2

2a

3

3a

3b

F. C. Moore lith.

W. BUCKLER del.

West, Newman & Co. imp.

PLATE LII.

Saturnia carpini.

1, larva before last moult; 1 *a*, 1 *b*, 1 *c*, larvæ after last moult.

Cilix spinula.

2, 2 *a*, larvæ after last moult.

Platypteryx lacertinaria.

3, 3 *a*, larvæ after last moult.

Drepana sicula.

4, larva before last moult; 4 *a*, 4 *b*, larvæ after last moult; 4 *c*, pupa.

See pp. 66—72.

1 a

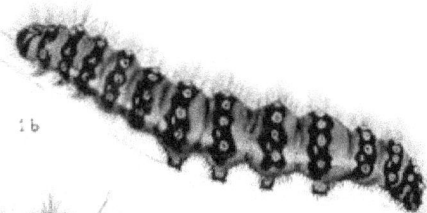

1 b

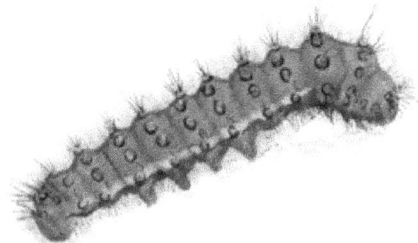

1 c

2

3

3 a

2 a

4

4 a

4 b

4 c

PLATE LIII.

Drepana falcataria.

1, 1 *a*, larvæ after last moult.

Drepana hamula.

2, 2 *a*, larvæ after last moult.
See p. 73.

Drepana unguicula.

3, 3 *a*, larvæ after last moult.

Psyche nigricans.

4, larva after last moult ; 4 *a*, case of larva.

Psyche fusca.

5, larva after last moult ; 5 *a*, case of larva.

Fumea nitidella.

6, adult larva and case.

Heterogenea asellus.

7, 7 *a*, adult larvæ slightly magnified ; 7 *b*, adult larva natural size on beech leaf ; also the cocoon.
See pp. 73—76.

Limacodes testudo.

8, 8 *a*, adult larvæ.

Plate LIII

F.C Moore lith.

West, Newman & C° imp

W BUCKLER del.

RAY SOCIETY.

INSTITUTED 1844.

FOR THE PUBLICATION OF WORKS ON NATURAL HISTORY.

ANNUAL SUBSCRIPTION ONE GUINEA.

LIST

OF

COUNCIL, OFFICERS, LOCAL SECRETARIES, AND MEMBERS,

TOGETHER WITH THE

TITLES OF THE PUBLICATIONS OF THE SOCIETY

CORRECTED TO FEBRUARY, 1889.

LIST OF LOCAL SECRETARIES.

Birmingham	...	...	... W. R. Hughes, Esq.
Leeds ...	...	...	... L. C. Miall, Esq.
South Devon	...	..	... G. C. Bignell, Esq.
Norwich	...		... F. W. Harmer, Esq.
Warrington	...		T. G. Rylands, Esq.

LIST OF SUBSCRIBERS.*

Aberdeen, University of.

Adkin, R., Esq., F.E.S., Wellfield, Lingards road, Lewisham, S.E.

Adlard, J. E., Esq., Bartholomew close, E.C.

Alderson, Mrs., Park House, Worksop, Notts.

Allen, E. S., Esq., Shepherds Green, Chiselhurst.

Allman, Professor, F.R.S., &c., Ardmore, Parkstone, Poole, Dorset.

American Institute, New York.

Anderson, J., Esq., junr., Alre Villas, Chichester.

Andrews, Arthur, Esq., Newtown House, Blackrock, Dublin.

Angelin, Professor, Stockholm.

Argyll, Duke of, F.R.S., Argyll Lodge, Kensington, W.

Armstrong, Lord, F.R.S., The Minories, Newcastle-on-Tyne.

Army and Navy Club, 36, Pall Mall, S.W.

Ash, Rev. C. D., Brisbane, Queensland.

Asher, Messrs., 13, Bedford street, W.C.

Ashley, R., Esq., Pinehurst, Basset, Southampton.

Ashmolean Society, Oxford.

Asiatic Society of Bengal, 57, Park street, Calcutta (per Messrs. Trübner).

Athenæum Club, Pall Mall, S.W.

Aubrey, Rev. H. G. W., Rectory, Hale, Salisbury.

Ackland Museum.

Babington, Professor Charles C., M.A., F.R.S., Cambridge.

Baer, Herr J., Frankfort.

Baillière, Messrs., 20, King William street, W.C.

Baker, G. T., Esq., 16, Clarendon road, Edgbaston, Birmingham.

* The Subscribers are requested to inform the Secretary of *any errors or omissions* in this List, and of any delay in the transmission of the Yearly Volume.

Balfour, Prof. J. B., F.R.S., Botanic Gardens, Edinburgh.

Baltimore, Peabody Institute.

Bankes, E. R., Esq., M.A., The Rectory, Corfe Castle.

Barker, H. W., Esq., 148, Hollydale road, Queen's road, Peckham, S.E.

Barrett, C. G., Esq., F.E.S., Norfolk street, King's Lynn.

Bastian, Dr. H. C., F.R.S., F.L.S., 20, Queen Anne street, W.

Bath Microscopical Society, care of C. Terry, Esq., 6, Gay street, Bath.

Becker, Rev. W., Willow Green Cottage, Willow, Newark-on-Trent, Notts.

Belfast Society for Promoting Knowledge, Linen Hall, Belfast.

Belfast, Queen's College.

Bergen, Museum of, Norway.

Berlin Royal Library.

Bignell, G. C., Esq., F.E.S., 7, Clarence place, Stonehouse, Plymouth.

Billups, T. R., Esq., F.E.S., 20, Swiss villas, Coplestone road, Peckham, S.E.

Binks, J., Esq., Wakefield.

Bird, G. W., Esq., Hurley Lodge, Honor Oak Park, S.E.

Birmingham, Free Library.

Birmingham, Mason College.

Birmingham Old Library.

Blatch, W. G., Esq., Small Heath, Birmingham.

Blomefield, Rev. L., F.L.S., Z.S., 19, Belmont, Bath.

Bloomfield, Rev. E. N., M.A., F.E.S., Guestling, near Hastings.

Bond, F., Esq., 5, Fairfield Avenue, Staines, Middlesex.

Bostock, E., Esq., Stone, Staffordshire.

Bostock, F., Esq., jun., Northampton.

Boston Public Library, U.S.A.

Bourne, T. W., Esq., Kyrewood, Tenbury.

Bower, B.A., Esq., Langley, Eltham Road, Lee, S.E.

Bowyer, R. W., Esq., Haileybury College, Hertford.

Brabant, Monsieur E., Chateau de l'Alouette, Escaudœuvres, France.

Bradford Naturalists' Society, E. Naylor, Esq., *Hon. Sec.*, 48, Market place, Bradford.

Bradley, R. C., Esq., 10, Digbeth, Birmingham.

Brady, H. B., Esq., F.L.S., 15, Robert street, Adelphi, W.C.

Brady, W. E., Esq., 1, Queen street, Barnsley, Yorkshire.

Braithwaite, Dr. R., F.L.S., The Ferns, Clapham rise, S.W.

Brevoort, Dr. J. Carson, New York.

Briggs, C. A., Esq., F.E.S., 55, Lincoln's Inn Fields, W.C.
Briggs, T. H., Esq., 55, Lincoln's Inn Fields, W.C.
Bright, P. M., Esq., Roccabruna, Bournemouth.
Brighton and Sussex Natural History Society, Brighton.
Bristol Microscopical Society, Dr. Harrison, *Hon. Sec.*, Fairland Lodge,
    Guthrie road, Clifton.
Bristowe, B. A., Esq., Durlstone, Champion hill, S.E.
Brockholes, Mrs. J. Fitzherbert, Clifton hill, Garstang, Lancashire.
Brodrick, W., Esq., Little hill, Chudleigh, South Devon.
Brook, Geo., Esq., jun., F.L.S., Fernbrook, Huddersfield.
Browell, E. M., Esq., Feltham, Middlesex.
Browne, Dr. Henry, Woodheys, Heaton Mersey, Manchester.
Buckmaster, Rev. C. J., Hindley Vicarage, Wigan.
Buckton, G. B., Esq., F.R.S., Weycombe, Haslemere, Surrey.
Burn, Dr. W. B., Beechwood, Balham road, Upper Tooting, S.W.
Burney, Rev. H., Wavendon Rectory, Woburn.

Cambridge, Rev. O. P., Bloxworth Rectory, Wareham.
Cambridge Entomological Society, per J. Copper, Esq., Hon. Sec.,
    Trinity College, Cambridge.
Cambridge, University Library.
Cambridge, University Museum of Zoology.
Cambridge, Downing College.
Cambridge, Gonville and Caius College.
Cambridge, St. Catharine's College.
Cambridge, Sidney-Sussex College.
Cambridge, Trinity College.
Campbell, F. M., Esq., F.L.S., Rose hill, Hoddesdon.
Canterbury, Philosophical Institute of, New Zealand.
Capper, S. J., Esq., F.L.S., Huyton Park, Huyton, near Liverpool.
Capron, Dr. E., F.E.S., Shiere, near Guildford, Surrey.
Carpenter, Dr. A., Duppas House, Croydon, S.
Carus, Dr. Victor, Leipsic.
Cash, W., Esq., F.G.S., F.L.S., F.R.M.S., Halifax, Yorkshire.
Chapman, E., Esq., Frewen Hall, Oxford.
Chapman, T. A., Esq., Firbank, Hereford.
Cheltenham Permanent Library, Cheltenham.
Chicago Library, Chicago.
Chichester and West Sussex Natural History Society, per C. T. Halstead,
    Esq., Hon. Treas., Chichester.

Christiania, University of.

Christy, W. M., Esq., Watergate, Emsworth, Hants.

Church, Dr. W. S., 130, Harley Street, W.

Cincinnati Public Library.

City of London Entomological Society, Albion Hall, London Wall, E.C.

Clark, J. A., Esq., M.P.S.G.B., L.D.S., F.E.S., The Broadway, London Fields, Hackney, E.

Cleland, Professor, 2, The College, Glasgow.

Colman, Jeremiah J., Esq., M.P., Carrow House, Norwich.

Cooper, Colonel E. H., 42, Portman square, W.

Cooper, Sir Daniel, Bart., 6, De Vere gardens, Kensington Palace, W.

Coppin, John, Esq., Bingfield House, by Corbridge-on-Tyne, R.S.O.

Cork, Queen's College, Cork.

Cornwall, Royal Institution of, Truro.

Crallan, G. E., Esq., Cambridgeshire Asylum, Fulbourn, near Cambridge.

Craven, Alfred E., Esq., 65, St. George's road, S.W.

Cregoe, J. P., Esq., Tredinick, Bodmin.

Cresswell, Mrs. R., Teignmouth, Devon.

Crisp, F., Esq., B.A., LL.B., V.P. and Treas. L. S., 6, Old Jewry, E.C.

Croft, R. Benyon, Esq., R.N., F.L.S., Farnham Hall, Ware, Herts.

Crowley, Philip, Esq., F.L.S., Wadden House, Croydon, S.

Cruickshank, Alexander, Esq., LL.D., 20, Rose street, Aberdeen.

Dale, C. W., Esq., F.E.S., Glanville Wootton, Sherborne, Dorset.

Daltry, Rev. T. W., M.A., F.L.S., Madeley Vicarage, Newcastle, Staffordshire.

Dawson, Sir J. W., F.R.S., F.G.S., M'Gill College, Montreal.

Dawson, W. G., Esq., Plumstead Common road, Plumstead, S.E.

Decie, Miss A. Prescott, Bockleton Court, Tenbury.

Devon and Exeter Institution, Exeter.

Devonshire, Duke of, F.R.S., 78, Piccadilly, W.

Dickinson, Wm., Esq., 3, Whitehall place, S.W.

Dobree, N. T., Esq., Beverley, Yorkshire.

Dohrn, Dr. Anton, Stazione Zoologica, Naples.

Douglas, W. D. R., Esq., F.L.S., Orchardton, Castle Douglas, N.B.

Downing, J. W., Esq., 59, Lupus street, St. George's square, S.W.

Dowsett, A., Esq., F.E.S., Castle Hill House, Reading.

Dublin, National Library.

Dublin Royal College of Science.
Dublin, Royal College of Surgeons.
Dublin, Royal Irish Academy.
Dublin, Trinity College.
Dublin, Hon. Society of King's Inn.
Ducie, Earl of, F.R.S., F.G.S., 16, Portman square, W.
Dunning, J. W., Esq., M.A., F.L.S., 12, Old square, Lincoln's Inn, W.C.

East Kent Natural History Society, Canterbury.
Edinburgh, Library of University of.
Edinburgh, Museum of Science and Art.
Edinburgh, Royal College of Physicians.
Edinburgh, Royal Society of.
Edinburgh, Royal Physical Society, 40, Castle street, Edinburgh.
Edwards, S., Esq., F.E.S., Kidbrooke Lodge, Blackheath, S.E.
Edwards, W. H., Esq., Coalburgh, West Virginia, United States.
Elisha, Geo., Esq., F.E.S., 122, Shepherdess Walk, City road.
Ellison, F. E., Esq., 3, Devon road, Fishponds, Bristol.
Ellison, S. T., Esq., 2, Balhousie street, Perth, N.B.
Elphinstone, H. W., Esq., F.L.S., 2, Stone Buildings, Lincoln's Inn, W.C.
England, Bank of, Library, London, E.C.
England, Royal College of Surgeons of, Lincoln's-inn-fields, W.C.
Essex Field Club, per A. P. Wire, Esq., Buckhurst hill, Essex.
Evans, H. A., Esq., United Services College, Westward Ho, Bideford, N. Devon.

Fenn, C., Esq., Eversden House, Burnt Ash Hill, Lee, S.E.
Ffarington, Miss M. H., Worden Hall, near Preston.
Fitch, E. A., Esq., F.L.S., Brick House, Maldon, Essex.
Fitch, Fred., Esq., F.R.G.S., Hadleigh House, Highbury New Park, N.
Flemyng, Rev. W. W., M.A., Coolfin House, Portlaw, Co. Waterford.
Fletcher, W. H. B., Esq., F.E.S., 6, The Steyne, Worthing, Sussex.
Flower, W. H., Esq., F.R.S., British Museum (Natural History), S.W.
Foster, C., Esq., Thorpe, Norwich.
Fraser, F. J., Esq., 16, Furnival Inn, E.C.
Freeman, F. F., Esq., F.E.S., 8, Leigham terrace, Plymouth.

Friedlander & Son, Messrs., Berlin.
Fuller, Rev. A., M.A., F.E.S., Pallant, near Chichester.

Galton, Sir Douglas, F.R.S., 12, Chester street, Grosvenor place, S.W.
Gardner, J., Esq., 8, Friar terrace, Hartlepool.
Gatty, C. H., Esq., M.A.,F.L.S., Felbridge place, East Grinstead, Sussex.
Geological Society, London, W.
Geological Survey of India, Calcutta, per Messrs. Trübner.
George, Frederick, Esq., Fairholme, Torquay.
Gibson, Mrs. G. S., Esq., Hill House, Saffron Walden, Essex.
Glasgow Natural History Society, 207, Bath street, Glasgow.
Glasgow, Philosophical Society of.
Glasgow, University of.
Godman, F. D., Esq., F.R.S., 10, Chandos street, Cavendish square,
    W., and South Lodge, Horsham.
Goldthwait, O. C., Esq., 2, Grove Villas, Grove road, Walthamstow.
Goode, J. F., Esq., 3, Regent place, Birmingham.
Gordon, Rev. George, LL.D., Manse of Birnie, by Elgin, N.B.
Göttingen, University of.
Green, R. Y., Esq., 11, Lovaine crescent, Newcastle-on-Tyne.
Grieve, Dr. J., F.R.S.E., F.L.S., care of W. L. Buchanan Esq., 212,
    St. Vincent street, Glasgow.
Grut, Ferdinand, Esq., F.L.S., 9, Newcomen street, Southwark, S.E.
Günther, Dr., F.R.S., British Museum (Natural History), Cromwell
    road, South Kensington, S.W.

Hackney Microscopical and Natural History Society, per J. A. Clark,
    Esq., Treasurer, 48, The Broadway, London fields, Hackney, E.
Haeckel, Professor, Jena, Prussia.
Hailstone, Edward, Esq., F.S.A., Walton Hall, Wakefield.
Hall, A. E., Esq., Norbury, Sheffield.
Hancock, John, Esq., Newcastle-on-Tyne.
Harbottle, A., Esq., 76, Mandle road, South Stockton.
Harley, Dr. J., F.L.S., 9, Stratford place, W.
Harmer, Sidney F., Esq., B.Sc., King's College, Cambridge.
Harris, Edw., Esq., F.G.S., Rydal Villa, Longton grove, Upper
    Sydenham.
Harris, J. T., Esq., F.E.S., Burton Bank, Burton-on-Trent.

Harrison, F., Esq., Junior United Service Club, Charles street, S.W.
Harvard College, Cambridge, U.S.A.
Havers, J. C., Esq., Joyce Grove, Nettlebed, Henley-on-Thames.
Hawker, H. G., Esq., Burleigh, Devonport.
Hawkshaw, J. C., Esq., 33, Great George street, Westminster, S.W.
Hepburn, Sir T. B., Bart., Smeaton, Preston Kirk, N.B.
Hertfordshire Natural History Society and Field Club, Watford.
Hickling, G. H., Esq., Mudie's Select Library, W.C.
Hicks, Dr. John B., F.R.S., 24, George street, Hanover square, W.
Hillier, J. T., Esq., 4, Chapel place, Ramsgate.
Hilton, James, Esq., 60, Montagu square, W.
Hinchliff, Miss Katharine M., Worlington House, Instow.
Holdsworth, E. W. H., Esq., F.L.S., 84, Clifton hill, Abbey road, N.W.
Hooker, Sir J. D., C.B., M.D., F.R.S., Sunningdale, Berks.
Hope, G. P., Esq., Upminster Hall, near Romford.
Hopkinson, John, Esq., F.L.S., F.G.S., The Grange, St. Alban's,
    Herts.
Horley, W. L., Esq., Stanboroughs, Hoddeston.
Houghton, Rev. W., F.L.S., Preston Rectory, Wellington, Salop.
Hovenden, F., Esq., F.L.S., Glenlea, Thurlow Park, Dulwich, S.E.
Howden, Dr. J. C., Sunnyside, Montrose.
Huddersfield Naturalists' Society, A. W. Whiteley, Esq., Hon. Sec.,
    Westgate, Huddersfield.
Hughes, W. R., Esq., F.L.S., *Local Secretary*, Wood House, Hands-
    wood, Birmingham.
Hughes, W. Rathbone, Esq., 3, Princes Gate East, Princes Park,
    Liverpool.
Hull Subscription Library.
Hunt, John, Esq., Milton of Campsie, Glasgow.
Hutchinson, Miss E., Grantsfield, Kimbolton, Leominster.
Huxley, Professor T. H., F.R.S., Science Schools, South Kensington.

Indian Museum, Calcutta.

James, H. B., Esq., F.Z.S., R.A.S., Valparaiso, Chili, care of Frank
    James, Esq., Aldridge, near Walsall, Staffordshire.
Janson, E. W., Esq., F.E.S., 35, Little Russell street, Bloomsbury.
Jenner, Charles, Esq., Easter Duddingsten Lodge, Portobello, Edin-
    burgh.

Jones, Albert H., Esq., Shrublands, Eltham.

Jordan, Dr. R. C. R., 35, Harborne road, Edgbaston, Birmingham.

Kane, W. F. de V., Esq., M.R.I.A., F.E.S., Sloperton Lodge, Kingstown, Co. Dublin.

Keays, F. Lovell, Esq., F.L.S., 26, Charles street, St. James', S.W.

Kenderdine, F., Esq., Morningside, Old Trafford, Manchester.

Kenrick, G. H., Esq., Whetstone, Somerset road, Edgbaston, Birmingham.

Keys, J. H., Esq., 8, Princes street, Plymouth.

Kilmarnock Library, Kilmarnock.

King, A., Esq., Aspley Guise, Woburn, Bedfordshire.

Kirby, W. F., Esq., 5, Burlington gardens, Chiswick, W.

Langdale, H. M., Esq., Thornycroft, Compton, Petersfield, Hants.

Laver, H., Esq., F.L.S., Colchester.

Laxton, H., Esq., 41, Harpur street, Bedford.

Leeds Philosophical and Literary Society.

Leicester, Alfred, Esq., Hollymount, Albert road, Birkdale, near Southport.

Leicester Free Library, Wellington street, Leicester.

Leipzig, University of.

Lemann, F.C., Esq., M.E.S., Blackfriars House, Plymouth.

Lidstone, W. G., Esq., 79, Union street, Plymouth.

Linnean Society, Burlington House, Piccadilly, W.

Lister, Arthur, Esq., F.L.S., Leytonstone.

Liverpool, Athenæum.

Liverpool Free Library.

Liverpool Library, Lyceum, Liverpool.

Liverpool Medical Institution.

Liverpool Microscopical Society.

Liverpool, Royal Institution.

Lloyd, A., Esq., F.E.S., The Dome, Bognor, Sussex.

Lochbuie Marine Institute, Lochbuie, N.B.

Longstaff, G. B., Esq., M.D., Southfield Grange, West Hill road, Wandsworth, S.W.

London Institution, Finsbury circus, E.C.

London Library, 12, St. James's square, S.W.

Lovén, Professor, Stockholm.

Lubbock, Sir J., Bart., M.P., F.L.S., R.S., *President*, 15, Lombard street, E.C.

Lupton, H., Esq., Lyndhurst, North Grange road, Headingley.

Marlborough College Natural History Society, Marlborough.

McGill, H. J., Esq., Aldenham Grammar School, Elstree, Herts.

McGregor, Rev. J., West Green, Culross, Dunfermlime, N.B.

McIntosh, Prof. W. C., M.D., F.R.S., 2, Abbotsford crescent, St. Andrew's, N.B.

McLachlan, R., Esq., F.R.S., West View, Clarendon road, Lewisham, S.E.

McMillan, W. S., Esq., 17, Temple street, Liverpool.

Maclagan, Sir Douglas, M.D., F.R.S.E., 28, Heriot row, Edinburgh.

Maclaine, M. G., of Lochbuie, Isle of Mull.

Madras Government Museum, Madras.

Major, Charles, Esq., Red Lion Wharf, 69, Upper Thames street, E.C.

Manchester Free Public Library.

Manchester Literary and Philosophical Society.

Mansel-Pleydell, J., Esq., F.L.S., Whatcombe, Blandford.

Marshall, A. E., Esq., Waldersea, Beckenham.

Martin, G. M., Esq., Red Hill Lodge, Compton, Wolverhampton.

Mason, P. B., Esq., F.L.S., Burton-on-Trent.

Mathew, G. F., Esq., R.N., F.L.S., Z.S., Lee House, Dovercourt, near Harwich, Essex.

Mathews, W., Esq., M.A., F.G.S., 60, Harborne road, Birmingham.

Matthews, C., Esq., F.E.S., Erme Wood, Ivy Bridge, S. Devon.

Meiklejohn, Dr. J. W. S., F.L.S., 105, Holland road, Kensington, W.

Melbourne Public Library.

Mennell, H. T., Esq., F.L.S., 10, St. Dunstan's buildings, Idol lane, E.C.

Michael, A. D., Esq., F.L.S., Cadogan Mansions, Sloane square, S.W.

Microscopical Society, Royal, King's College, Strand, London.

Miller, J. C., Esq., Lynmouth House, Langley road, Elmers End, Beckenham, Kent, S.E.

Mitchell Library, the, Glasgow.

Mivart, Prof. St. George J., F.R.S., Chilworth, Guildford.

Moore, Mrs. E. T., Holmfield, Oakholme road, Sheffield.

Moseley, Sir T., Rolleston Hall, Burton-on-Trent.

Munich Royal Library, Munich.
Murdock, J. Barclay, Esq., F.R.Ph.S.E., Barclay, Langside, Glasgow.

Neave, B. W., Esq., Lyndhurst, Queen's road, Brownswood park, N.
Newcastle Literary and Philosophical Society, Newcastle-upon-Tyne.
Newman, T. P., Esq., 54, Hatton garden, E.C.
Noble, Capt. Jesmond Dene House, Newcastle-on-Tyne.
Noble, Wilson, Esq., 43, Warrior square, St. Leonard's-on-Sea.
Norfolk and Norwich Library, Norwich.
Norman, Rev. A. Merle, M.A., F.L.S., Burnmoor Rectory, Fencehouses, Durham.
Nottingham Free Library.
Nottingham Naturalists' Society, per W. H. Kay, Esq., Hon. Sec., Gresham Chambers, Nottingham.

Oldfield, G. W., Esq., M.A., F.L.S., 6, South Bank terrace, Stratford road, Kensington, W.
Owens College, Manchester.
Oxford, Magdalen College.

Paisley Philosophical Institute, Paisley.
Paris National Library, per Messrs. Longmans.
Parke, Geo. H., Esq., Infield Lodge, Barrow-in-Furness.
Parker, W. K., Esq., F.R.S., Crowland, Trinity road, Upper Tooting, S.W.
Pascoe, F. P., Esq., F.L.S., 1, Burlington road, Westbourne Park, W.
Pearce, W. G., 187, Caledonian road, King's Cross, N.
Peckover, Algernon, Esq., F.L.S., Wisbeach.
Peel Park Library, Salford, Lancashire.
Penny, Rev. C. W., Wellington College, Wokingham.
Penzance Public Library.
Perthshire Society of Natural Science, Museum, Tay street, Perth.
Phené, J. S., Esq., LL.D., F.S.A., 5, Carlton terrace, Oakley street, S.W.
Philadelphia Academy of Natural Sciences, U.S.A.
Pierce, F. Nelson, Esq., 143, Smithdown lane, Liverpool.
Plymouth Institution, Athenæum, Plymouth.
Pole-Carew, Miss C. L., 3, South place, Rutland gate, S.W., and Antony, Torpoint, Devonport.

Pode, E. D. Y., Esq., 133, Fentimans road, S.W.
Porritt, G. T., Esq., F.L.S., Greenfield House, Huddersfield.
Poulton, E. B., Esq., Wykeham House, Oxford.
Power, H., Esq., 37A, Great Cumberland place, Hyde Park, W.
Preston Free Public Library.
Pye-Smith, Dr. P. H., 54, Harley street, Cavendish square, W.

Quekett Microscopical Club, University College, W.C.

Radcliffe Library, Oxford.
Radford, D., Esq., Mount Tavy, Tavistock, Devon.
Ramsay, Sir Andrew C., F.R.S., 7, Victoria terrace, Beaumaris.
Rashleigh, J., Esq., Menabilly, Par Station, Cornwall.
Reader, Thomas, Esq., 39, Paternoster row, E.C.
Reading Microscopical Society, 110, Oxford road, Reading.
Reynell, Miss, 8, Henrietta Street, Dublin.
Ripon, Marquis of, F.R.S., F.L.S., 1, Carlton gardens, S.W.
Robinson, Rev. F., The Rectory, Castle Eden, Co. Durham.
Robinson, Isaac, Esq., The Wash, Hertford.
Robson, J. E., Esq., 15, Northgate, Hartlepool.
Roper, F. C. S., Esq., F.L.S., F.G.S., Palgrave House, Eastbourne.
Rose, Geo., Esq., Queen street, Barnsley.
Ross, J. G., Esq., Bathampton Lodge, Bathampton, Bath.
Rothery, H. C., Esq., M.A., F.L.S., 94, Gloucester terrace, Hyde Park, W.
Royal Institution, Albemarle street, W.
Royal Medical and Chirurgical Society, 53, Berners street, W.
Royal Society, Burlington House, London, W.
Rowe, J. B., Esq., F.L.S., Plympton Lodge, Plympton, S. Devon.
Rowland-Brown, H., Esq., jun., Oxhey grove, Stanmore.
Rylands, T. G., Esq., F.L.S., *Local Secretary*, High Fields, Thelwall,
    near Warrington.

Salter, Dr. S. J. A., F.R.S., *Treasurer*, Basingfield, near Basingstoke,
    Hants.
Salvin, Osbert, Esq., F.R.S., 10, Chandos street, Cavendish square.
Samson and Wallin, Messrs., London.
Sanders, Alfred, Esq., F.L.S., Milton, Sittingbourne, Kent.

Sanford, W. A., Esq., F.G.S., Nynehead Court, near Wellington, Somersetshire.

Science and Art Department, South Kensington.

Sclater, P. L., Esq., M.A., Ph.D., F.L.S., R.S., 11, Hanover square, W.

Sharpus, F. W., Esq., 30, Compton road, Islington, N.

Sheffield Literary and Philosophical Society.

Sheldon, Dr. T. S., Cheshire County Asylum, Macclesfield.

Shillitoe, B., Esq., 2, Frederick place, Old Jewry, E.C.

Sinclair, R. S., Esq., 16, Annfield terrace W., Parkhill, Glasgow.

Sion College Library, Victoria Embankment, W.C.

Slack, H. I., Esq., F.G.S., Ashdown Cottage, Forest row, Sussex.

Sladen, Rev. C. A., Burghclere, Newbury.

Slatter, T. J., Esq., F.G.S., Evesham.

Smith, Basil Woodd, Esq., F.R.A.S., Branch hill, Hampstead, N.W.

Smith, F. W., Esq., F.E.S., Hollywood, Lewisham hill, S.E.

Smith, S. P., Esq., F.E.S., 22, Rylett road, Shepherd's Bush, W.

Somersetshire Archæological and Natural History Society, Taunton.

Sotheran, Messrs., 136, Strand, W.C.

South London Entomological Society, The Bridge House, London Bridge.

South London Microscopical Club, care of J. Guardia, Esq., Helston House, Rozel road, Clapham, S.W.

South, R., Esq., F.E.S., 12, Abbey gardens, St. John's Wood, N.W.

Southport Free Library.

Spicer, Messrs., Brothers, 19, New Bridge street, Blackfriars, E.C.

St. Andrew's University Library, St. Andrew's.

Stainton, H. T., Esq., F.R.S., L.S., Mountsfield, Lewisham, S.E.

Stebbing, Rev. T. R. R., Ephraim Lodge, The Common, Tunbridge Wells.

Stedman, A., Esq., M.R.C.S., L.S.A., L.M., The Croft, Great Bookham, Leatherhead.

Stephenson, J. W., Esq., Equitable Assurance Office, Mansion-house street, E.C.

Stewart, Prof. C., F.L.S., Royal College of Surgeons, Lincoln's Inn Fields, W.C.

Stockholm Royal Academy, Stockholm.

Strasbourgh University Library.

Stubbins, J., Esq., F.G.S., R.M.S., Inglebank, Far Headingly, Leeds.

Sunderland Subscription Library.

Swanston, W., Esq., F.G.S., 50, King street, Belfast.

Thompson, J. C., Esq., F.L.S., R.M.S., Woodstock, Waverley road' Liverpool.
Thornewell, Rev. C. F., The Soho, Burton-on-Trent.
Tomlinson, J. H., Esq., 7, Kirkgate, Newark.
Toronto, University of, Canada.
Torquay Natural History Society, Museum, Babbacombe road, Torquay.
Townsend, F., Esq., M.A., Honington Hall, Shipston-on-Stour.
Trimble, Mrs. James, 2, Clarendon road, Southsea, Portsmouth.
Trübner & Co., Messrs., London.
Tugwell, W. H., Esq., 6, Lewisham road, Greenwich, S.E.
Turner, Professor W., F.R.S.E., University of Edinburgh.
Tyler, Captain Charles, F.L.S., F.G.S., Elberton, New West End, Hampstead, N.W.

University College, London.
Upsala, University of, Sweden.
Vass, M., Leipzig.
Vicars, John, Esq., 8, St. Alban's square, Bootle, Liverpool.
Vicary, William, Esq., The Priory, Colleton crescent, Exeter.
Vinen, Dr. E. Hart, F.L.S., 22, Gordon road, Ealing, W.

Waldegrave, Earl, 13, Montagu place, Montagu square, W.
Walker, Alfred O., Esq., Chester.
Walker, Rev. Dr. F. A., F.L.S., Duis Mallard, Cricklewood, N.W.
Walsingham, Thomas de Grey, Lord, M.A., F.L.S., Z.S., Merton Hall, Thetford, Norfolk.
Warburgh, J. C., Esq., 8, Porchester terrace, W.
Warden, Dr. Charles, Greenhurst, 31, Newall street, Birmingham.
Warrington Museum and Library, Warrington.
Warwickshire Natural History Society, Warwick.
Washington Library of Congress, U.S.A.
Watkinson Library, Harford, Con., U.S.A.
Webb, S., Esq., Maidstone House, Dover.
Weir, J. J., Esq., F.L.S., Chirbury, Copers Cope road, Beckenham, Kent.
Wells, J. R., Esq., 20, Fitzroy street, Fitzroy square, W.C.
Wesley, E. F., Esq., A.K.C., 28, Essex street, Strand, W.C.
West Kent Natural History Society, Lewisham, S.E.

Wheeler, F. D., Esq., Paragon House School, Norwich.

Whittle, F. G., Esq., 2, Cambridge terrace, Lupus street, S.W.

Wilson, Owen, Esq., F.E.S., Cwmffrwd, Carmarthen.

Wiltshire, Rev. Professor T., M.A., F.L.S., Treas. G.S., *Secretary*, 25, Granville park, Lewisham, London, S.E.

Wollaston, G. H., Esq., 4, College road, Clifton, near Bristol.

Wood, J. H., Esq., M.B., Tarrington, Ledbury.

Woodd, B. T., Esq., Conyngham Hall, Knaresborough, Yorkshire.

Wright, Professor E. P., F.L.S., Trinity College, Dublin.

Yale College, New Haven, U.S.

York Philosophical Society, York.

Zoological Society, 11, Hanover square, W.

www.ingramcontent.com/pod-product-compliance
Lightning Source LLC
Chambersburg PA
CBHW020535270326
41927CB00006B/587